Natural Language Processing with Flair

A practical guide to understanding and solving NLP problems with Flair

Tadej Magajna

BIRMINGHAM—MUMBAI

Natural Language Processing with Flair

Copyright © 2022 Packt Publishing

Publishing Product Manager: Ali Abidi
Senior Editor: Nathanya Dias
Content Development Editor: Nazia Shaikh
Technical Editor: Devanshi Ayare
Copy Editor: Safis Editing
Project Coordinator: Aparna Ravikumar Nair
Proofreader: Safis Editing
Indexer: Sejal Dsilva
Production Designer: Aparna Bhagat
Marketing Coordinators: Abeer Dawe, Shifa Ansari

First published: April 2022

Production reference: 1210422

Published by Packt Publishing Ltd.
Livery Place
35 Livery Street
Birmingham
B3 2PB, UK.

ISBN 978-1-80107-231-1

www.packt.com

To my girlfriend and my family, who supported me throughout the entire process of writing this book. Special thanks to Prof. Dr. Alan Akbik, the original author of Flair, for help and for making the framework available to the public.

– Tadej Magajna

Contributors

About the author

Tadej Magajna is a former lead machine learning engineer, former data scientist, and now a software engineer at Microsoft. He currently works in a team responsible for language model training and building language packs for keyboards such as Microsoft SwiftKey. He has a master's degree in computer science. He started his career as a 15-year-old at a local media company as a web developer and progressed toward more complex problems. He has tackled problems such as NLP market research, public transport bus and train capacity forecasting, and finally, language model training in his current role. Currently, he is based in his hometown of Ljubljana, Slovenia.

About the reviewers

After studying engineering and physics, **Pascal Tartarin** held various senior management positions (sales, marketing, and R&D) mainly in the chemical and medical device industries in Europe and Asia Pacific. A keen lifelong learner, he later developed his expertise in data science and NLP. Over the last 6 years, he has built applications in the field of business (forecasts, CRM) based on ML, deep learning, NLP, and more recently, knowledge graphs (Neo4j).

Amardeep Kumar has sound experience in Natural Language Processing and machine learning systems. He has industry experience of 2 years working on different technical stacks in NLP and software development. Amardeep regularly contributes to open source software development as a developer, mentor, and reviewer in Google Summer of Code programs. He has published research papers for top-tier NLP-proc conferences, such as ACL and EMNLP.

Table of Contents

3
Embeddings in Flair

4
Sequence Tagging

Part 2: Deep Dive into Flair – Training Custom Models

5
Training Sequence Labeling Models

6

Hyperparameter Optimization in Flair

7

Train Your Own Embeddings

10

Hands-On Exercise – Building a Trading Bot with Flair

Preface

Natural Language Processing (**NLP**) is currently one of the fastest-growing subfields of AI where new advances or changes to NLP frameworks happen daily. Anyone new to the field will find it difficult to know where to start, which ideas are outdated, which ones are still relevant, and which are soon to become state of the art. Similarly, people with a solid theoretical NLP background aiming to gain practical experience in NLP will find it difficult to choose the right NLP library or framework given the oversupply of different options currently available in the open source community. As someone who has used Flair both professionally as well as for personal projects, I would find it hard to recommend any other framework that is as intuitive, as relevant, and as packed with ready-to-use, state-of-the-art models as Flair.

The most interesting thing about Flair may not be its simplicity, performance, or ease of use. Instead, it has to do with how Flair was originally designed. The first release was never intended to serve as a full-fledged NLP framework. The first release of Flair (v0.1), at heart, was merely a tool that served as a real-world implementation of Flair embeddings – the underlying concept that gives Flair sequence labeling models their amazing performance. It wasn't until later Flair versions that the library introduced Transformer models, text classification models, and other tools that you would normally expect from an established NLP framework.

Anyone aiming to build real-world NLP solutions will need to juggle between spending time learning more about NLP theory and investing effort in choosing the right engineering approach. This book will help you find the right balance between both. It will arm you with just the right amount of theoretical foundation to help understand the underlying NLP concepts as well as help you gain enough engineering knowledge and experience so that you can use Flair proficiently.

Who this book is for

This book is for anyone eager to learn about NLP through one of the most beginner-friendly, yet powerful, Python NLP libraries out there. The audience includes software engineering students, developers, data scientists, and anyone who is transitioning into NLP and is interested in learning about practical approaches to solving problems with Flair. This book, however, is not recommended for readers aiming to get an in-depth theoretical understanding of the mathematics behind NLP. Beginner-level knowledge of Python is recommended to get the most out of this book.

The book is also an opportunity for anyone hoping to start contributing to the open source community. You could fix bugs, create new tutorials, expand documentation, add support for new languages by building pre-trained models, or simply open new GitHub issues with suggestions about improving Flair. The knowledge gained by reading and doing the exercises in this book will help you considerably on your way to becoming an active Flair contributor.

What this book covers

Chapter 1, Introduction to Flair, provides a quick overview of NLP and its basic problems and techniques. The chapter then introduces Flair and shows how to set up your local environment.

Chapter 2, Flair Base Types, introduces Flair's basic syntax, its typical classes, and methods.

Chapter 3, Embeddings in Flair, explains the concept behind word and document embeddings and their role in NLP. It describes all the different types of embeddings available in Flair.

Chapter 4, Sequence Tagging, describes sequence tagging and its subtypes, such as named entity recognition and part-of-speech tagging. This chapter also demonstrates how to use pre-trained sequence taggers in Flair.

Chapter 5, Training Sequence Labeling Models, explains how to train, save, and use custom sequence tagging models in Flair.

Chapter 6, Hyperparameter Optimization in Flair, shows the importance of using the right hyperparameters for model training. It introduces hyperparameter optimization tools in Python and explains how to perform hyperparameter optimization in Flair.

Chapter 7, Train Your Own Embeddings, explains how to train custom embeddings in Flair and how to leverage different evaluation techniques for measuring success.

Chapter 8, Text Classification in Flair, introduces the problem of text classification. This chapter demonstrates how to use pre-trained models as well as how to train custom classifiers. It also introduces a novel approach to text classification called TARS.

Chapter 9, Deploying and Using Models in Production, talks about the challenges of deploying and using NLP models in production. This chapter demonstrates how to set up custom minimum viable product NLP services and how to host Flair models on the Hugging Face models hub.

Chapter 10, Hands-On Exercise – Building a Trading Bot with Flair, solves a real-world problem as part of a hands-on exercise by building a trading bot with Flair.

To get the most out of this book

To run the code in this book, you will need a Python environment version 3.6 or later. It is highly recommended you run the code as part of a Jupyter notebook session, which will allow you to experiment and try out different variations of the original code more easily. To bypass the local setup, you can also run the code in a hosted Jupyter notebook environment such as Google Colab. You will also need the following Python packages:

Python packages needed to run the code in this book	Operating system requirements
Flair 0.11	Windows, macOS, or Linux
Flask 2.0.3	Windows, macOS, or Linux

Detailed instructions about how to set up your local environment and install the Python packages can be found in *Chapter 1, Introduction to Flair*.

If you are using the digital version of this book, we advise you to type the code yourself or access the code from the book's GitHub repository (a link is available in the next section). Doing so will help you avoid any potential errors related to the copying and pasting of code.

Download the example code files

You can download the example code files for this book from GitHub at `https://github.com/PacktPublishing/Natural-Language-Processing-with-Flair/`. If there's an update to the code, it will be updated in the GitHub repository.

We also have other code bundles from our rich catalog of books and videos available at `https://github.com/PacktPublishing/`. Check them out!

Download the color images

We also provide a PDF file that has color images of the screenshots and diagrams used in this book. You can download it here: `https://static.packt-cdn.com/downloads/9781801072311_ColorImages.pdf`.

Conventions used

There are a number of text conventions used throughout this book.

`Code in text`: Indicates code words in the text, variable names, class names, object methods, and properties.

A block of code is set as follows:

```
from flair.data import Sentence
from flair.tokenization import SpaceTokenizer

tokenizer = SpaceTokenizer()
s = Sentence('Some nice text.', use_tokenizer=tokenizer)

print(s)
```

When we wish to draw your attention to a particular part of a code block, the relevant lines or items are set in bold:

```
from flair.data import Sentence

sentence = Sentence('A short sentence')
sentence.get_token(1).add_label('manual-pos', 'DT')

print(sentence)
```

Any command-line input or output is written as follows:

```
$ python3 --version
$ pip3 install flair==0.11
```

Keyword: Indicates a new term, an important name, or a concept of great significance; for example, **Hugging Face** transformer models.

> **Tips or Important Notes**
> Appear like this.

Get in touch

Feedback from our readers is always welcome.

General feedback: If you have questions about any aspect of this book, email us at customercare@packtpub.com and mention the book title in the subject of your message.

Errata: Although we have taken every care to ensure the accuracy of our content, mistakes do happen. If you have found a mistake in this book, we would be grateful if you would report this to us. Please visit www.packtpub.com/support/errata and fill in the form.

Piracy: If you come across any illegal copies of our works in any form on the internet, we would be grateful if you would provide us with the location address or website name. Please contact us at copyright@packt.com with a link to the material.

If you are interested in becoming an author: If there is a topic that you have expertise in and you are interested in either writing or contributing to a book, please visit authors.packtpub.com.

Share Your Thoughts

Once you've read *Natural Language Processing with Flair*, we'd love to hear your thoughts! Scan the QR code below to go straight to the Amazon review page for this book and share your feedback.

https://packt.link/r/1801072310

Your review is important to us and the tech community and will help us make sure we're delivering excellent quality content.

Part 1: Understanding and Solving NLP with Flair

In this part, you will learn the basics of NLP and get an overview of the Flair framework. You will set up your environment, install Flair, and explore its basic features. You will learn how to extract knowledge from embeddings and use pre-trained sequence labeling models in Flair.

This part comprises the following chapters:

1
Introduction to Flair

There are few **Natural Language Processing (NLP)** frameworks out there as easy to learn and as easy to work with as **Flair**. Packed with pre-trained models, excellent documentation, and readable syntax, it provides a gentle learning curve for NLP researchers who are not necessarily skilled in coding; software engineers with poor theoretical foundations; students and graduates; as well as individuals with no prior knowledge simply interested in the topic. But before diving straight into coding, some background about the motivation behind Flair, the basic NLP concepts, and the different approaches to how you can set up your local environment may help you on your journey toward becoming a Flair NLP expert.

In Flair's official GitHub README, the framework is described as:

"A very simple framework for state-of-the-art Natural Language Processing"

This description will raise a few eyebrows. NLP researchers will immediately be interested in knowing what specific tasks the framework achieves its state-of-the-art results in. Engineers will be intrigued by the *very simple* label, but will wonder what steps are required to get up and running and what environments it can be used in. And those who are not knowledgeable in NLP will wonder whether they will be able to grasp the knowledge required to understand the problems Flair is trying to solve.

In this chapter, we will be answering all of these questions by covering the basic NLP concepts and terminology, providing an overview of Flair, and setting up our development environment with the help of the following sections:

- A brief introduction to NLP

- What is Flair?

- Getting ready

Technical requirements

To get started, you will need a development environment with Python 3.6+. Platform-specific instructions for installing Python can be found at `https://docs.python-guide.org/starting/installation/`.

You will not require a GPU-equipped development machine, though having one will significantly speed up some of the training-related exercises described later in the book.

You will require access to a command line. On Linux and macOS, simply start the **Terminal** application. On Windows, press *Windows + R* to open the **Run** box, type `cmd` and then click **OK**.

Flair's official GitHub repository is available via the following link: `https://github.com/flairNLP/flair`. In this chapter we will install Flair version 0.11.

The code examples covered in this chapter are found in this book's official GitHub repository in the following Jupyter notebook: `https://github.com/PacktPublishing/Natural-Language-Processing-with-Flair/tree/main/Chapter01`.

A brief introduction to NLP

Before diving straight into what Flair is capable of and how to leverage its features, we will be going through a brief introduction to NLP to provide some context for readers who are not familiar with all the NLP techniques and tasks solved by Flair. NLP is a branch of artificial intelligence, linguistics, and software engineering that helps machines understand human language. When we humans read a sentence, our brains immediately make sense of many seemingly trivial problems such as the following:

- Is the sentence written in a language I understand?

- How can the sentence be split into words?

- What is the relationship between the words?

- What are the meanings of the individual words?

- Is this a question or an answer?

- Which part-of-speech categories are the words assigned to?

- What is the abstract meaning of the sentence?

The human brain is excellent at solving these problems conjointly and often seamlessly, leaving us unaware that we made sense of all of these things simply by reading a sentence.

Even now, machines are still not as good as humans at solving all these problems at once. Therefore, to teach machines to understand human language, we have to split understanding of natural language into a set of smaller, machine-intelligible tasks that allow us to get answers to these questions one by one.

In this section, you will find a list of some important NLP tasks with emphasis on the tasks supported by Flair.

Tokenization

Tokenization is the process of breaking down a sentence or a document into meaningful units called **tokens**. A token can be a paragraph, a sentence, a collocation, or just a word.

For example, a word tokenizer would split the sentence *Learning to use Flair* into a list of tokens as `["Learning", "to", "use", "Flair"]`.

Tokenization has to adhere to language-specific rules and is rarely a trivial task to solve. For example, with unspaced languages where word boundaries aren't defined with spaces, it's very difficult to determine where one word ends and the next one starts. Well-defined token boundaries are a prerequisite for most NLP tasks that aim to process words, collocations, or sentences including the following tasks explained in this chapter.

Text vectorization

Text vectorization is a process of transforming words, sentences, or documents in their written form into a numerical representation understandable to machines.

One of the simplest forms of text vectorization is **one-hot encoding**. It maps words to binary vectors of length equal to the number of words in the dictionary. All elements of the vector are 0 apart from the element that represents the word, which is set to 1 – hence the name *one-hot*.

For example, take the following dictionary:

- Cat
- Dog
- Goat

The word **cat** would be the first word in our dictionary and its one-hot encoding would be `[1, 0, 0]`. The word **dog** would be the second word in our dictionary and its one-hot encoding would be `[0, 1, 0]`. And the word **goat** would be the third and last word in our dictionary and its one-hot encoding would be `[0, 0, 1]`.

This approach, however, suffers from the problem of high dimensionality as the length of this vector grows linearly with the number of words in the dictionary. It also doesn't capture any semantic meaning of the word. To counter this problem, most modern state-of-the-art approaches use representations called word or document embeddings. Each embedding is usually a fixed-length vector consisting of real numbers. While the numbers will at first seem unintelligible to a human, in some cases, some vector dimensions may represent some abstract property of the word – for example, a dimension of a word-embedding vector could represent the general (positive or negative) sentiment of the word. Given two or more embeddings, we will be able to compute the similarity or distance between them using a distance measure called **cosine similarity**. With many modern NLP solutions, including **Flair**, embeddings are used as the underlying input representation for higher-level NLP tasks such as named entity recognition.

One of the main problems with early word embedding approaches was that words with multiple meanings (polysemic words) were limited to a single and constant embedding representation. One of the solutions to this problem in **Flair** is the use of contextual string embeddings where words are contextualized by their surrounding text, meaning that they will have a different representation given a different surrounding text.

Named entity recognition

Named entity recognition (**NER**) is an NLP task or technique that identifies named entities in a text and tags them with their corresponding categories. Named entity categories include, but aren't limited to, *places*, *person names*, *brands*, *time expressions*, and *monetary values*.

The following figure illustrates NER using colored backgrounds and tags associated with the words:

Berkeley **ORG**   University **ORG**   is located in   Berkeley **LOC** .

Figure 1.1 – Visualization of NER tagging

In the previous example, we can see that three entities were identified and tagged. The first and third tags are particularly interesting because they both represent the same word, **Berkeley**, yet the first one clearly refers to an organization whereas the second one refers to a geographic location. The human brain is excellent at distinguishing between different entity types based on context and is able to do so almost seamlessly, whereas machines have struggled with it for decades. Recent advancements in contextual string embeddings, an essential part of Flair, made a huge leap forward in solving that.

Word-sense disambiguation

Word-Sense Disambiguation (**WSD**) is an NLP technique concerned with identifying the intended sense of a given word with multiple meanings.

For example, take the given sentence:

George tried to return to Berlin to return his hat.

WSD would aim to identify the sense of the first use of the word *return*, referring to the act of giving something back, and the sense of the second *return*, referring to the act of going back to the same place.

Part-of-speech tagging

Part-of-Speech (**POS**) tagging is a technique closely related to both WSD and NER that aims to tag the words as corresponding to a particular part of speech such as nouns, verbs, adjectives adverbs, and so on.

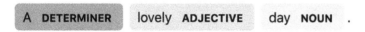

Figure 1.2 – Visualization of POS tagging

Actual POS taggers provide a lot more information with the tags than simply associating the words with noun/verb/adjective categories. For example, the Penn Treebank Project corpus, one of the most widely used NER corpora, distinguishes between 36 different types of POS tags.

Chunking

Another NLP technique closely related to POS tagging is **chunking**. Unlike **parts of speech (POS)**, where we identify individual POS, in chunking we identify complete short phrases such as noun phrases. In *Figure 1.2*, the phrase **A lovely day** can be considered a chunk as it is a noun phrase, and in its relationship to other words works the same way as a noun.

Stemming and lemmatization

Stemming and **lemmatization** are two closely related text normalization techniques used in NLP to reduce the words to their common base forms. For example, the word *play* is the base word of the words *playing*, *played* and *plays*.

The simpler of the two techniques, **stemming**, simply accomplishes this by cutting off the ends or beginnings of words. This simple solution often works, but is not foolproof. For example, the word *ladies* can never be transformed into the word *lady* by stemming only. We therefore need a technique that understands the POS category of a word and takes into account its context. This technique is called *lemmatization*. The process of lemmatization can be demonstrated using the following example.

Take the following sentence:

```
this meeting was exhausting
```

Lemmatization reduces the previous sentence to the following:

```
this meeting be exhaust
```

It reduces the word was to be and the word exhausting to exhaust. Also note that the word meeting is used as a noun and it is therefore mapped to the same word meeting, whereas if the word meeting was used as a verb, it would be reduced to meet.

A popular and easy-to-use library for performing lemmatization with Python is **spaCy**. Its models are trained on large corpora and are able to distinguish between different POS, yielding impressive results.

Text classification

Text classification is an NLP technique used to assign a text or a document to one or more classes or document types. Practical uses for text classification include spam filtering, language identification, sentiment analysis, and programming language identification from syntax.

Having covered the basic NLP concepts and terminology, we can now move on to understanding what Flair is and how it manages to solve NLP tasks with state-of-the-art results.

Introducing Flair

Flair is a powerful NLP framework published as a Python package. It provides a simple interface that is friendly, easy to use, and caters to people from various backgrounds including those with little prior knowledge in programming. It is published under the MIT License, which is one of the most permissive free software licenses.

Flair as an NLP framework comes with a variety of tools and uses. It can be defined in the following ways:

- *It is an NLP framework used in NLP research for producing models that achieve state-of-the-art results across many NLP tasks such as POS tagging, NER, and chunking across several languages and datasets.* In Flair's GitHub repository, you will find step-by-step instructions on how to reproduce these results.

- *It is a tool for training, validating, and distributing NER, POS tagging, chunking, word sense disambiguation, and text classification models.* It features tools that help ease the training and validation processes such as the automatic corpora downloading tool, and tools that facilitate model tuning such as the hyperparameter optimization tool. It supports a growing number of languages.

- *It is a tool for downloading and using state-of-the-art pre-trained models.* The models are downloaded seamlessly, meaning that they will be automatically downloaded the first time you use them and will remain stored for future use.

- *It is a platform for the proposed state-of-the-art Flair embeddings.* The state-of-the-art results Flair achieves in many NLP tasks can by and large be attributed to its proposed Flair contextual string embeddings described in more detail in the paper *Contextual String Embeddings for Sequence Labeling.* The author refers to them as "the secret sauce" of Flair.

- *It is an NLP framework for working with biomedical data.* A special section of Flair is dedicated solely to working with biomedical data and features a set of pretrained models that achieve state-of-the-art results, as well as a number of corpora and comprehensive documentation on how to train custom biomedical tagging models.

- *It is a great practical introduction to NLP.* Flair's extensive online documentation, simple interface, inclusive support for a large number of languages, and its ability to perform a lot of the tasks on non-GPU-equipped machines all make it an excellent entry point for someone aiming to learn about NLP through practical hands-on experimentation.

Setting up the development environment

Now that you have a basic understanding of features offered by the framework, as well as an understanding of the basic NLP concepts, you are now ready to move to the next step of setting up your development environment for Flair.

To be able to follow the instructions in this section, first make sure you have Python 3.6+ installed on your device as described in the *Technical requirements* section.

Creating the virtual environment

In Python, it's generally good practice to install packages in virtual environments so that the project dependencies you are currently working on will not affect your global Python dependencies or other projects you may work on in the future.

We will use the `venv` tool that is part of the Python Standard Library and requires no installation. To create a virtual environment, simply create a new directory, move into it, then run the following command:

```
$ python3 -m venv learning-flair
```

Then, to activate the virtual environment on Linux or macOS, run the following:

```
$ source learning-flair/bin/activate
```

If you are running Windows, run the following:

```
$ learning-flair\Scripts\activate.bat
```

Your command line should become prefixed with `(learning-flair) $` and your virtual environment is now active.

Installing a published version of Flair in a virtual environment

You should now be ready to install Flair version 0.11 with this single command:

```
(learning-flair) $ pip install flair==0.11
```

The installation should now commence and finish within a minute or so depending on the speed of your internet connection.

You can verify the installation by running the following command, which will display a list of package properties including its version:

```
(learning-flair) $ pip show flair
Name: flair
Version: 0.11
Summary: A very simple framework for state-of-the-art NLP
Home-page: https://github.com/flairNLP/flair
...
```

A command output like the preceding indicates the package has been successfully installed in your virtual environment.

Installing directly from the GitHub repository (optional)

In some cases, the features we aim to make use of in Flair may already be implemented in a branch on GitHub, but those changes may not yet be released as part of a Python package published on PyPI. We can install Flair with those features directly from the Git repository branch.

For example, here is how you can install Flair from the master branch:

```
(learning-flair) $ git clone https://github.com/flairNLP/flair.
git
(learning-flair) $ cd flair
(learning-flair) $ git checkout master
(learning-flair) $ pip install .
```

> **Important note**
> Installing code from non-reviewed branches can introduce unreliable or unsafe code. When installing Flair from development branches, make sure the code you are installing comes from a trusted source. Also note that the future versions of Flair (versions larger than 0.11) may not be compatible with the code snippets found in this book.

Replace the term master with any other branch name to install the package from a branch of your choice.

Running code that uses Flair

Running code that makes use of the Flair Python package is no different from running any other type of Python code.

The recommended way for you to run the code snippets in this book is to execute them as code cells in a Jupyter notebook, which you can install and run as follows:

```
(learning-flair) $ pip install notebook
(learning-flair) $ jupyter notebook
```

You can then create a new Python 3 notebook and run your first Flair script to verify the package is imported successfully:

```
import flair
print(flair.__version__)
```

After executing, the preceding code should print out the version of Flair you are currently using, indicating that the Flair package has been imported successfully and you are ready to start.

Summary

In this chapter, you became familiar with the basic NLP terminology and tasks. As you learn about Flair, you will often come across terms such as tokenization, NER, and POS, and the knowledge gained in this chapter will help you understand what they mean.

You also now understand where Flair sits in the NLP space, what problems it's solving and which fields it excels in. Finally, you've learned how to install Flair inside your virtual environment either from a PyPI package or a Git branch. You are now ready to start coding with Flair!

In the upcoming chapter, we will be covering basic syntax and the basic objects in Flair, known as base types.

2
Flair Base Types

A good place to start with any NLP framework is getting comfortable with its basic objects and methods used frequently throughout the code. In Flair, the first step is getting familiar with its **base types**. These are the basic objects that are used for defining sentences or text fragments and forming tokens through tokenization.

One of the main challenges NLP is struggling with today is its support for underrepresented languages. Most state-of-the-art prebuilt NLP models are usually built only for some of the most spoken languages, while failing to provide support for the roughly 7,000 other languages spoken on the planet today. While Flair stands out with its excellent language coverage and its work on multilingual embeddings, it's still far from supporting all the possible languages in areas such as *corpora availability*, *tokenization methods*, and *prebuilt sequence tagging models*. To form tokens for a language with special tokenization rules currently not supported by Flair, you will need to implement your own tokenizer. Luckily, doing so in Flair couldn't be easier.

In this chapter, you will learn about the Sentence object, which is used for representing sentences or text fragments, and the Token base type, which is used for representing words in tokenized sentences. You will learn how to implement custom tokenizers and how to use them. You will also learn about the Corpus object in Flair and how to use its helper functions to load and interact with corpora and datasets.

This chapter covers the following topics:

- Sentence and Token objects
- Using custom tokenizers
- Corpus objects

Technical requirements

For this chapter, you will need a **Python** environment with **Flair version 0.11** installed and a stable internet connection since we will be downloading corpora.

The code examples in this chapter can be found in this book's official GitHub repository, in the following Jupyter notebook: `https://github.com/PacktPublishing/Natural-Language-Processing-with-Flair/tree/main/Chapter02`.

Sentence and Token objects

Sentence and **Token** objects can be regarded as the most common objects in **Flair** syntax. The former is used for representing sentences or any other text fragments such as paragraphs. They are essentially lists of `Token` objects and their corresponding label assignments.

If you are wondering what objects, classes, and methods mean, this simply suggests that you're not particularly familiar with **object-oriented programming (OOP)**. Luckily, **OOP** is super easy to get a grasp of in Python. In **OOP**, pieces of code that store and/or process data are organized into blueprints called classes. An example of a class could be the `Word` class, which can store and process a single word. Classes in OOP can include several procedures called methods. An example of a `Word` class method could be `get_length()`, which would simply return a word's length. Classes can also contain a special type of method called a constructor, which gets called when the class is instantiated. The result of instantiating a class is an object. For example, calling `word_1 = Word('potato')` would call the constructor method with `'potato'` as the first argument value, return a `Word` object, and store it as `word_1`. We can then use this variable to call this object's methods, for example, `word_1.get_length()`.

This brief introduction to OOP should give you enough understanding to be able to dive straight into Flair's arguably most fundamental entity – the `Sentence` class.

Understanding the Sentence class

The Sentence class constructor accepts the following parameters:

- text: Original string (sentence) or a list of string tokens (words).
- use_tokenizer: Tokenizer to use. If True, SegTokTokenizer will be used. If False, SpaceTokenizer will be used instead. If a Tokenizer object is provided, it will be used as the custom tokenizer.
- language_code: The language code that's used for some embeddings (optional).
- start_position: The start char offset of the sentence in the document (optional).

For example, let's create a simple Sentence object and examine its string representation behavior:

```
from flair.data import Sentence

sentence = Sentence('Some nice text.')
print(sentence, len(sentence))
```

The preceding script should print out the following:

```
Sentence: "Some nice text ." 4
```

The preceding output indicates that the sentence has been split into four tokens, one of which is a full stop. The separate . token is a direct result of not providing the use_tokenizer parameter, which, if left blank, defaults to using SegTokTokenizer.

Understanding the Token class

The Token class represents a single entity such as a word in a tokenized sentence.

Each token can have zero, one or multiple tags. For example, Flair allows using the same Sentence object for both POS tagging and NER. This means that a single token can in theory have both a named entity label as well as a part-of-speech label.

The Token object's corresponding text can easily be obtained by referring to its Token.text property.

We now have a vague understanding of how the two base types work. But to be able to fully understand how base types work, we need to dive deeper into tokenization techniques in Flair.

Tokenization in Flair

Depending on your needs and the type of language used, you may choose from any of the following tokenizers available in Flair:

- `SpaceTokenizer()`: A simple tokenizer that splits text by the space character. Note that this tokenizer splits by the space character (`" "`) only. It does not split by characters that are included in Python's `string.whitespace`, which is often used in methods for splitting or formatting text. Whitespace (along with the space character) also includes the new line and tab characters, which are not taken into account with `SpaceTokenizer`.

- `SpacyTokenizer(model)`: A tokenizer that uses a commonly used NLP library called **spaCy**. It accepts a parameter called `model` that can either be a string representing a **spaCy** model (for example, **en_core_web_sm**) or a spaCy `Language` object.

- `SciSpacyTokenizer()`: A tokenizer that uses spaCy's **en_core_sci_sm** model, extended by special characters such `(, )`, and `-`. These serve as additional token separators.

- `SegtokTokenizer()`: A tokenizer that uses a Flair dependency called **segtok** that focuses on splitting (Indo-European) text.

- `JapaneseTokenizer(tokenizer, sudachi_mode)`: A tokenizer class that uses the **konoha** library and provides a selection of tokenizers for the Japanese language. The `tokenizer` parameter can be either `mecab`, `janome`, or `sudachi`.

Now that we have a good understanding of the tokenizers available in Flair, let's try to use the `Sentence` object with a non-default tokenizer:

```python
from flair.data import Sentence
from flair.tokenization import SpaceTokenizer

tokenizer = SpaceTokenizer()
s = Sentence('Some nice text.', use_tokenizer=tokenizer)

# getting the string representation using magic method __str__
print(s, len(s))
```

The preceding script, which uses `SpaceTokenizer`, should print out the following:

```
Sentence: "Some nice text." 3
```

This, unlike `SegtokTokenizer`, treats the `text.` string as a single token due to splitting only on the space character.

In the next section, we will look at how to extract information from the `Sentence` and `Token` objects and how to tag them with additional data.

Sentence and Token object helper methods

The `Sentence` object has several helpful methods and properties. The most straightforward one, and one that we already used in the preceding example, is the implementation of the `__str__` magic method, which returns a string representation of the object, for example, when trying to print it out.

To be able to fully understand the format of the string representation of the `Sentence` object, we need to tag at least one token. To do that, we will use the `get_token(n)` method to get the `Token` object for the *nth* token, and then use the `add_label(label_type, value)` method to assign a label to the token:

```
from flair.data import Sentence

sentence = Sentence('A short sentence')
sentence.get_token(1).add_label('manual-pos', 'DT')
print(sentence)
```

The preceding script will print out the following:

```
Sentence: "A short sentence" → ["A"/DT]
```

The first part of the printed-out string displays the entire sentence in its original form, whereas the second part of the sentence displays the tokens with their corresponding tags. In this code example, we manually tagged the first token to be able to test out the string representation of the object, whereas in Flair, tokens will usually be labeled by sequence taggers instead.

A similar string representation of the `Sentence` object can also be directly obtained by calling the `to_tagged_string()` method.

> **Important Note**
>
> The `get_token(n)` method uses 1-based indexing, meaning that `n=1` will retrieve the first token, whereas `n=0` will return None. A 0-based indexing alternative to using `get_token(n)` would be to use the indexer operator. For example, to get the first token for a `Sentence` object called `sentence`, we can simply call `sentence[0]`.

Tokens in a tokenized `Sentence` object can also be obtained by using its `__iter__()` magic function, which allows us to get all the tokens by simply iterating the object. We can, for example, do that by using a `for` loop.

Let's run the following script using the same `sentence` variable from the preceding script:

```
for token in sentence:
    print(token)
```

The script will print out the following:

```
Token[0]: "A" → DT (1.0)
Token[1]: "short"
Token[2]: "sentence"
```

This allows us to get a clear visual representation of how our original string was tokenized.

When working with the `Sentence` object, it is important to understand the behavior of using the `len()` method with the object. It will return the number of tokens, not the length of the original string. This means that calling `len(sentence)` using the variable in the preceding script would return 3. If you wish to get the length of the actual sentence (that is, the original text) you will need to run `len(sentence.to_original_text())`.

If none of the tokenizers described here work for you, Flair has you covered. All you need to do is implement your own custom tokenizer and pass it into Flair. Let's learn how.

Using custom tokenizers

While Flair ships with several tokenizers that support the most commonly spoken languages, it is entirely possible you will be working with a language that uses tokenization rules currently not covered by Flair. Luckily, Flair offers a simple interface that allows us to implement our tokenizers or use third-party libraries.

Using the TokenizerWrapper class

The `TokenizerWrapper` class provides an easy interface for building custom tokenizers. To build one, you simply need to instantiate the class by passing the `tokenizer_func` parameter. The parameter is a function that receives the entire sentence text as input and returns a list of token strings.

As an exercise, let's try to implement a custom tokenizer that splits the text into characters. This tokenizer will treat every character as a token:

```
from flair.data import Token
from flair.tokenization import TokenizerWrapper


def char_splitter(sentence):
    return list(sentence)


char_tokenizer = TokenizerWrapper(char_splitter)
```

In the preceding code, we implemented the `space_splitter` function, which simply splits the sentence into a list of characters. Finally, we created the new tokenizer object by instantiating the `TokenizerWrapper` class with our new function.

Now, we can test our tokenizer by using it with a `Sentence` object and printing out the generated tokens:

```
from flair.data import Sentence


text = "Good day."
sentence = Sentence(text, use_tokenizer=char_tokenizer)


for token in sentence:
    print(token)
```

The resulting script will print out the following:

```
Token[0]: "G"
Token[1]: "o"
Token[2]: "o"
Token[3]: "d"
Token[4]: " "
Token[5]: "d"
```

```
Token[6]: "a"
Token[7]: "y"
Token[8]: "."
```

The printed text indicates that our tokenizer performs as expected and that it is compatible with the `Sentence` object.

Important Note on Tokenizers

Tokenization rules for languages are often complex and include many edge cases. Therefore, it is always recommended to use third-party tokenizers from established NLP packages as opposed to implementing your own.

Now that we've covered tokenization and the `Sentence` and `Token` base types, it's time to move on and dive into understanding the `Corpus` object.

Understanding the Corpus object

The `Corpus` object is the main object that stores corpora in memory in Flair. Each `Corpus` is a collection of three datasets that behave like lists of `Sentence` objects. These datasets can be accessed via the following properties:

- The **train** property, which contains the dataset that will be used for training models.
- The **test** property, which contains a dataset that's independent of the train dataset. It is used for model validation.
- The **dev** property, which contains the dataset that's used for hyperparameter tuning.

These three datasets ideally contain data from the same data source and follow the same probability distribution.

An example corpus object can be obtained by loading one of Flair's prepared datasets:

```
from flair import datasets

corpus = datasets.UD_ENGLISH()
```

The corpus summary can be obtained by simply printing out the object:

```
print(corpus)
```

This should print out the following:

```
Corpus: 12543 train + 2002 dev + 2077 test sentences
```

The train dataset of the corpus can be obtained by simply calling the `train` property:

```
train_dataset = corpus.train
```

Then, the nth sentence of the dataset can be obtained by using the indexer operator. For example, the 101st sentence can be obtained by using the following code:

```
sentence = train_dataset[100]
print(sentence)
```

The preceding script will print out the 101st sentence in the train dataset.

The object also ships with several helper methods. One of the most widely used is the `downsample(percentage)` method, which reduces the size of the corpus, where `percentage` is a value between 0 and 1 that determines the size of the new corpus.

> **Note**
> The `downsample(percentage)` method modifies and returns the original Corpus object. This means that calling it several times downsamples the corpus multiple times.

For example, if we wanted to downsample our corpora to 1%, we would use the following code:

```
corpus.downsample(0.01)


print(corpus)
```

This will print out the following:

```
Corpus: 125 train + 20 dev + 21 test sentences
```

This is 1% of the size of the original corpus.

Flair ships with several techniques and instances of the `Corpus` object that allow us to load data from CSV files and read our own sequence labeling datasets. It offers a vast selection of prepared datasets, all of which will be covered in more detail later in the book.

Summary

In this chapter, we covered Flair's base types, such as the `Sentence` and `Token` objects, explained how to initialize and use them, and tried out some of their basic helper methods. This should allow us to handle, transform, and understand data in Flair more easily as we move toward more complex topics. We also covered using custom tokenizers in Flair and implemented our own character-based tokenizer. Finally, we scraped the surface of what Flair's datasets and the `Corpus` objects can do. We learned how to load corpora and datasets, assess their size, extract, and read individual sentences, and downsample entire datasets.

We are now familiar enough with the syntax, basic objects and helper methods to be able to move on to Flair's most powerful NLP technique – sequence tagging. We will cover this in the next chapter.

3
Embeddings in Flair

Embeddings are a core concept of **Natural Language Processing (NLP)** tasks in **Flair**. Understanding how different embeddings work, what types are available and knowing what type to use in what scenario is important in becoming a seasoned Flair user.

While the framework offers a large selection of different embedding types, there is one that is absolutely essential and plays a pivotal role in achieving Flair's excellent performance on NLP tasks. This type of embeddings, called Flair embeddings, is referred to by the main author as *the secret sauce of Flair*. Understanding the ins and outs of Flair embeddings will help you better understand the success behind Flair across so many NLP tasks and allow you to choose the right embeddings when training custom models.

In this chapter, we will learn about the different embedding types, how they work, and how to use the individual embedding techniques in Flair for producing word, sentence, and document embeddings. After reading this chapter, you will get a deeper understanding of what embeddings to use in what situation. We will demonstrate how you can use arithmetic operations on word embeddings to understand the meaning and relationships between words using the "king is to man as queen is to woman" analogy. We will learn how to form new embedding types by combining multiple existing embeddings using stacked embeddings. We will also cover document embeddings and how to use them to produce embeddings for entire documents.

The framework offers a range of embeddings that can be classified into various categories: *classic word embeddings*, *Flair embeddings*, *stacked embeddings*, *document embeddings*, and *other embeddings*. These categories are covered in the following topics:

- Understanding word embeddings
- Flair embeddings
- Stacked embeddings
- Document embeddings
- Other embeddings in Flair

Technical requirements

All of the Python libraries used in this chapter are direct Flair dependencies of Flair version 0.11 and will require no special setup, assuming Flair is already installed on your machine. The code examples covered in this chapter are found in this book's official GitHub repository in the following Jupyter notebook: `https://github.com/PacktPublishing/Natural-Language-Processing-with-Flair/tree/main/Chapter03`.

Understanding word embeddings

Word embeddings are machine-interpretable representations of words such that embeddings of word pairs with similar meanings will have similar embeddings and words with dissimilar meanings will have vastly different embeddings.

In the first chapter, where we covered the basics of embeddings, we loosely defined embeddings as vector representations of a particular character, word, sentence, paragraph, or text document. These vectors are often made up of hundreds or thousands of real numbers. Each position in this vector is referred to as a dimension.

By now, you've probably wondered how we can tell whether two word embeddings are similar or different from each other. There are several metrics, such as cosine similarity, Euclidean distance, and Jaccard distance, that try to quantify this. **Cosine similarity** is usually the most commonly used method.

Cosine similarity

Given two embedding vectors, *A* and *B*, cosine similarity is defined as follows:

$$similarity(\mathbf{A}, \mathbf{B}) = \frac{\mathbf{A} \cdot \mathbf{B}}{\|\mathbf{A}\| \|\mathbf{B}\|} = \frac{\sum_{i=1}^{n} A_i B_i}{\sqrt{\sum_{i=1}^{n} A_i^2} \sqrt{\sum_{i=1}^{n} B_i^2}}$$

Here, the numerator (the top part of the fraction) is a dot product of the two vectors and the denominator (the bottom part of the fraction) is the product of magnitudes of each vector.

The resulting similarity score is a value between *-1* and *1*. A value of *-1* indicates that the vectors are the exact opposite of each other. A value of *0* indicates that the vectors are orthogonal, meaning there is no correlation. A value of *1* indicates the vectors are similar.

For example, if we compared two embedding vectors representing two customers' music tastes where they rated different songs, we would have the following:

- *similarity*(**A**, **B**) = 1; means that the customers rated the same exact songs exactly the same.

- *similarity*(**A**, **B**) = −1; means that the customers rated the same songs in the opposite way.

- *similarity*(**A**, **B**) = 0; means that there is no correlation between the customers' music tastes, meaning they probably agreed on some songs, but disagreed on other.

You should now have a high-level understanding of the role and value of embeddings in NLP. It's now time to put the theory into practice and get some hands-on experience with working with embeddings with Flair.

The "king – man ≈ queen – woman" analogy

Well-designed and trained embeddings with a high enough dimensionality not only represent or uniquely identify a certain word or a sentence but also contain abstract information about the meaning of the word and its semantic properties. Surprisingly enough, we can use arithmetic operations on word vectors to form new words. This means that, in theory, if we add two word vectors together, we should get a new word that is the sum of the meanings of both of those words. Similarly, if we subtract one embedding vector from another, we get a new embedding that contains the meaning of the first word minus the meaning of the second word. This brings us to the **king – man** ≈ **queen – woman** analogy.

If we take the word embedding for the word *king* and subtract the word *man* from that embedding vector, we should get an embedding for a word embedding representing some sort of a gender-neutral monarch. Similarly, if we take the word embedding for the word *queen* and subtract the word *woman* from it, we should also get a gender-neutral monarch word representation.

By using the preceding assumptions, we can come up with the following equation:

$$king - man \approx queen - woman$$

> **Important note**
>
> Note that we used the *approximately equal to* symbol. It means that even if this claim holds, the two embeddings will only be approximately equal. The reason for this is mainly that the word embeddings are not perfect representations of words and are limited by their dimensionality, training biases, and the completeness of training data. The relationships between two pairs of words are also very unlikely to be identical.

What this analogy is saying is that the relationship between the words *king* and *man* is very similar to the relationship between the words *queen* and *woman*. In other words, we could say the following:

king is to man as queen is to woman

In theory, we should be able to apply this concept to any two pairs of words that share similarly matching relationships. To take this hypothesis further, let's replace our monarchs, *man*, and *woman*, with word placeholders such as *A, B, C,* and *D*:

$$A - B \approx C - D$$

This allows us to transform this equation so that we can compute *D*:

$$D \approx B + C - A$$

In other words, say we have the following statement:

A is to B as C is to D

We can now compute the embedding for the word D given the embeddings for words A, B, and C. Once we get the embedding for the word D, finding its corresponding word is easy. We simply search embeddings of the entire vocabulary and find the word with the closest matching embedding.

It's now time we put our analogy solver to the test with Flair.

Implementation

We are going to implement an algorithm that, given the words A, B, and C, computes the word D, in such a way that the *A is to B as C is to D* analogy will hold. For example, when given the input *King is to man as queen is to...*, the algorithm will respond with *woman*. The algorithm will consist of the following steps:

1. Choosing and instantiating the embedding class we will be using in Flair

2. Obtaining the embeddings for words A, B, and C

3. Computing the approximate embedding for the word D using the $D = B + C - A$ formula

4. Obtaining embeddings for all English words in Flair

5. Finding the closest matching word

Choosing and initializing the embedding class

When choosing embedding types, it's important that we choose a technique that isn't contextual because we will be computing new words independently of the context we might find them in.

We will therefore choose a set of random classic word embeddings using the `crawl` ID. The `crawl` embeddings are English word embeddings trained over web crawl data. The embedding object can simply be initialized by running the following:

```
from flair.embeddings import WordEmbeddings
fasttext = WordEmbeddings('crawl')
```

The first time this code is executed, Flair will download the embedding model, which takes up some space and may take a few minutes to download fully.

To save us from having to download the model again, Flair will make sure to use the locally stored version next time we use it.

Once the embedding object is loaded, we can keep it in memory and reuse and pass it to our helper functions down the line to save execution time.

Obtaining the embeddings for words A, B, and C and computing the embedding D

As we learned in previous chapters, we can embed words in Flair by first wrapping them in the `Sentence` object. For example, given a `fasttext` object such as the one defined previously, our `Sentence` object can be defined with `sentence = Sentence('word')`. Then, we can embed the words in our `Sentence` object by running `fasttext.embed(sentence)`. We can also obtain the embedding for the first word in the sentence by calling `sentence[0].embedding`. This will return a **PyTorch** `torch.Tensor` object. It can be turned into a list of vector components by calling `sentence[0].embedding.tolist()`.

With this knowledge, we can implement a `compute_embedding_for_D(A, B, C, embedding)` function that receives our three words (A, B, and C) and an embedding object and returns the embedding vector of the computed word D.

The implementation of this method is as follows:

```python
from flair.data import Sentence


def compute_embedding_for_D(A, B, C, embedding):
    wordsABC_sentence = Sentence(' '.join([A, B, C]))
    embedding.embed(wordsABC_sentence)

    A_embedded = wordsABC_sentence[0].embedding
    B_embedded = wordsABC_sentence[1].embedding
    C_embedded = wordsABC_sentence[2].embedding

    D_embedding = B_embedded + C_embedded - A_embedded

    return D_embedding.tolist()
```

In the first part of this helper function, we define the `wordsABC_sentence` object, which contains a space-separated string containing our three input words, A, B, and C. We then proceed to embed all the words in that sentence and extract embeddings for each of our three input words and thus obtain our word embeddings called `A_embedded`, `B_embedded`, and `C_embedded`. Using our $D = B + C - A$ formula and Python's arithmetic operators, we then simply compute `D_embedding` and return it as a list.

To test our method, we can run the following:

```
from flair.embeddings import WordEmbeddings
fasttext = WordEmbeddings('crawl')
D = compute_embedding_for_D('king', 'man', 'queen', fasttext)

print(D)
```

This should print out our D embedding – a long list of numbers looking something like this:

```
[0.13569998741149902, -0.3856000006198883, -0.1022999957203865,
-0.13539999723434448, -0.01410001516342163,
-0.04699999839067459, ...]
```

Given our *King is to man as queen is to woman* analogy, the resulting embedding should be similar to the embedding of the word *woman*. To verify that, we need to first compute embeddings of our English vocabulary and then find the closest matching word.

Obtaining embeddings for all English words in Flair

To obtain a list of the words we can, to a reasonable extent, expect in the English language, we are going to extract a vocabulary from a large English corpus. This way, we are going to get a near-complete set of commonly used English words while getting rid of the rarely used words in English which we would be extremely unlikely to ever stumble upon, thus reducing our search space and speeding up our code.

We will then embed those words as part of a single sentence containing a space-delimited list of our vocabulary. This can only be done when using non-contextual word embeddings. If we were to use contextual string embeddings, we would have to define one Sentence object for each word separately.

We will implement the solution as the get_embedded_english_vocab(embedding) function, which receives the embedding object and returns a Sentence object containing our entire vocabulary as tokens along with their corresponding embeddings:

```
from flair import datasets
from flair.data import Sentence

def get_embedded_english_vocab(embedding):
    dataset = datasets.UD_ENGLISH()
    vocab_list = dataset.make_vocab_dictionary().get_items()
```

```
      vocab = Sentence(' '.join(vocab_list))
      embedding.embed(vocab)
      return vocab
```

The get_embedded_english_vocab(embedding) function uses Flair's UD_ ENGLISH Universal Dependency Treebank for English dataset. It concatenates the dataset's entire vocabulary into a space-separated string that is then tokenized and embedded as part of the Sentence object.

The preceding function can be tested by running the following:

```
from flair.embeddings import WordEmbeddings
fasttext = WordEmbeddings('crawl')


print(get_embedded_english_vocab(fasttext)[6].embedding)
```

Here, we initialize a word embedding class and pass it to the get_embedded_english_ vocab function. The snippet will then finally print out the embedding of the seventh word in our vocabulary.

Given that we now have access to embeddings of all the words we can reasonably expect in the English language, we can now compute the word closest to our computed D word embedding.

Finding the closest matching word

Once we've computed the D word embedding, we finally need to find the word that our embedding represents. Because the embedding is highly unlikely to be an exact match with a word from our vocabulary, we need to find the closest matching word using some sort of a similarity measure. We will use the cosine similarity. To find the closest matching word, we will simply compute the cosine similarity between D and every word in our vocabulary and choose the word with the maximum similarity score. This will be implemented as find_closest_matching_word(D, vocab, ABC), where D is our computed embedding, vocab is our embedded vocabulary, and ABC is a list of the words A, B, and C:

```
from sklearn.metrics.pairwise import cosine_similarity as sim


def find_closest_matching_word(D, vocab, ABC):
    max_match = -1
```

```
for word in vocab:
    match = sim([D], [word.embedding.tolist()])[0][0]
    if match > max_match and word.text not in ABC:
        max_match = match
        closest_matching_word = word.text
return closest_matching_word
```

The preceding code simply iterates through our `vocab` object and computes the similarity between the D embedding and each word in the `vocab` object. We then pick and return `closest_matching_word`, which is the solution to our analogy solver.

> **Tip**
>
> While our idea of using arithmetic operations to solve analogies makes sense in theory, it in practice often turns out that the results aren't as sensible as you would expect. What often happens is that the newly computed embedding is actually closest to one of our original words. Of course, we want D to be a new word and not one of the original words, A, B, or C. When this happens, we can simply ignore the closest matching word and choose the second, third, or fourth closest matching word. We do this by checking for `and word.text not in ABC` in the preceding `if` statement and thus avoid the result being one of the original words and increase our chances of producing a sensible result.

Having implemented all the required logic for our analogy solver, we can now write a simple helper function that integrates all the functions together and print out the solution to our analogy puzzle:

```
def A_is_to_B_as_C_is_to(A, B, C):
    fasttext = WordEmbeddings('crawl')
    result = compute_embedding_for_D(A, B, C, fasttext)
    vocab = get_embedded_english_vocab(fasttext)
    D = find_closest_matching_word(result, vocab, {A, B, C})

    print(f'{A} is to {B} as {C} is to {D}')
```

> **Speed optimizations**
>
> Note that in the preceding function, we instantiated the `WordEmbeddings`
> object only once and then passed it directly to our functions. This improves
> the performance of our solver. There are also many other optimizations we
> could have made, namely to do with how we found our closest matching word.
> Instead of iterating our vocabulary using a `for` loop and computing the cosine
> similarity for each word, we could have passed our entire vocabulary to the
> `cosine_similarity` method, which allows computing similarity for
> several embedding pairs at once.

It's now time to try our solution out!

Experimenting with the analogy solver

We can now try a few examples and see how our analogy solver performs in practice:

- `A_is_to_B_as_C_is_to("king", "man", "queen")` results in
  the following:

  ```
  king is to man as queen is to woman
  ```

- `A_is_to_B_as_C_is_to("do", "did", "go")` results in the following:

  ```
  do is to did as go is to went
  ```

- `A_is_to_B_as_C_is_to("bread", "baker", "meat")` results in
  the following:

  ```
  bread is to baker as meat is to butcher
  ```

- `A_is_to_B_as_C_is_to("London", "England", "Ljubljana")`
  results in the following:

  ```
  London is to England as Ljubljana is to Slovenia
  ```

- `A_is_to_B_as_C_is_to("life", "death", "beginning")` results in
  the following:

  ```
  life is to death as beginning is to ending
  ```

- `A_is_to_B_as_C_is_to("big", "bigger", "small")` results in
  the following:

  ```
  big is to bigger as small is to smaller
  ```

- A_is_to_B_as_C_is_to("man", "actor", "woman") results in the following:

```
man is to actor as woman is to actress
```

The preceding code represents some examples where our solver yields surprisingly good results. It is, however, fair to note that we applied a small cheat that makes sure the predicted word is never one of the original three words, which would have often been the case if our cheat was not there.

In this exercise, we gained some hands-on experience working with Flair and deepened our understanding and knowledge of word embeddings.

We will now cover a range of different embedding types supported in Flair.

Classic word embeddings in Flair

The simplest type of embedding in Flair that still contains semantic information about the word is called **classic word embeddings**. These embeddings are pre-trained and non-contextual and require no preprocessing. The non-contextual nature means one word always maps to one precomputed embedding regardless of the context. To use them, we simply instantiate a WordEmbeddings class by passing in the ID of the embedding of our choice. Then, we simply wrap our text into a Sentence object and call the embed(sentence) method on our WordEmbeddings class.

Let's obtain embeddings for a few random words using the FastText "crawl" embeddings:

```
from flair.data import Sentence
from flair.embeddings import WordEmbeddings

embedding = WordEmbeddings('crawl')
sentence = Sentence("one two three one")
embedding.embed(sentence)

for token in sentence:
    print(token.embedding)
```

The preceding code will print out embeddings of the words one, two, three, and one. Note that we printed the one token twice. Because WordEmbeddings class is non-contextual, the first and fourth tokens' embeddings are identical.

We can verify that with a simple check:

```
token1 = sentence[0]
token4 = sentence[3]

print(token1.embedding.tolist() == token4.embedding.tolist())
```

The preceding code will print out `True`, indicating that the two embeddings for the same word are equivalent.

> **Note**
> When loading an embedding for the first time, Flair will download the embedding model and store it locally. It will use the locally stored version for any subsequent calls.

Flair currently supports the following types of classic word embeddings:

- GloVe embeddings:
  - ID: `glove` (English)
- Komninos embeddings:
  - ID: `extvec` (English)
- Twitter embeddings:
  - ID: `twitter` (English)
- Turian embeddings (small):
  - ID: `turian` (English)
- FastText embeddings over web crawls:
  - ID: `crawl` (English)
- FastText embeddings over news and Wikipedia data:
  - ID: `ar` (Arabic), `bg` (Bulgarian), `ca` (Catalan), `cz` (Czech), `da` (Danish), `de` (German), `es` (Spanish), `en` (English), `eu` (Basque), `fa` (Persian), `fi` (Finnish), `fr` (French), `he` (Hebrew), `hi` (Hindi), `hr` (Croatian), `id` (Indonesian), `it` (Italian), `ja` (Japanese), `ko` (Korean), `nl` (Dutch), `no` (Norwegian), `pl` (Polish), `pt` (Portuguese), `ro` (Romanian), `ru` (Russian), `si` (Slovenian), `sk` (Slovak), `sr` (Serbian), `sv` (Swedish), `tr` (Turkish), `zh` (Chinese)

You should now have a good understanding of how to produce simple word embeddings in many languages using one of Flair's classic word embedding methods. We now have the foundation and understanding to move on to a slightly more complex, but also more interesting, topic, one that is regarded as the reason behind Flair's excellent sequence tagging performance, the so-called **secret sauce of Flair** – Flair embeddings.

Flair embeddings

Flair embeddings are a special type of contextual string embeddings that model words as a sequence of characters. They are the reason behind Flair's excellent sequence tagging performance and were essentially the motivation for the introduction of the Flair NLP framework. The *Contextual String Embeddings for Sequence Labeling* paper, an interesting and easy read written by the original creator of Flair, explains the inner workings of Flair embeddings brilliantly. But to grasp Flair embeddings from the perspective of an NLP engineer, we only need to understand their two properties: their contextuality and character-level sequence modeling.

Understanding the contextuality of Flair embeddings

The idea behind contextual string embeddings is that each word embedding should be defined by not only its syntactic-semantic meaning but also the context it appears in. What this means is that each word will have a different embedding for every context it appears in.

Each pre-trained Flair model offers a **forward** version and a **backward** version. Let's assume you are processing a language that, just like this book, uses the left-to-right script. The forward version takes into account the context that happens before the word – on the left-hand side. The backward version works in the opposite direction. It takes into account the context after the word – on the right-hand side of the word.

If this is true, then two same words that appear at the beginning of two different sentences should have identical forward embeddings, because their context is null. Let's test this out:

```
from flair.data import Sentence
from flair.embeddings import FlairEmbeddings


embedding = FlairEmbeddings('news-forward')
s1 = Sentence("nice shirt")
s2 = Sentence("nice pants")


embedding.embed(s1)
```

```
embedding.embed(s2)
```

```
print(s1[0].embedding.tolist() == s2[0].embedding.tolist())
```

In the preceding code, we instantiate the `news-forward` Flair embedding and compute an embedding for the word `nice` in the `nice shirt` context and the embedding for the word `nice` in the `nice pants` context. Because we are using a **forward** model, it takes into account only the context that occurs before the word. Also, because our word has no context on the left-hand side of its position in the sentence, the two embeddings are identical and the code assumes they are identical, and indeed prints out `True`.

Let's now put some context *before* our word in question. The differing context for the same word should result in two different embeddings:

```
s1 = Sentence("very nice shirt")
s2 = Sentence("pretty nice pants")
```

```
embedding.embed(s1)
embedding.embed(s2)
```

```
print(s1[1].embedding.tolist() == s2[1].embedding.tolist())
```

In the preceding code, we define two sentences that have differing preceding contexts for our word `nice` and because we use the `news-forward` embedding model, the two embeddings for the word `nice` should be different. We verify this in the last line, which assumes the two embeddings for the word `nice` are equivalent. The operation results in `False`. This indicates the forward type of Flair embeddings indeed takes into account the context on the left-hand side of the word.

Character-level sequence modeling in Flair embeddings

Usage of character-level sequence modeling in Flair embeddings has many benefits, but its most easily observable advantage is that it handles **out of vocabulary** (**OOV**) words exceptionally well. Some word embedding techniques, such as **GloVe**, offer no support for OOV words and return a vector of zeros for words not in their vocabulary. The OOV support not only offers good performance for words that never appeared in the original training set but also offers support for words that were mistyped.

Flair's excellent OOV word performance can be observed in the following experiment. We will obtain an embedding for a correctly typed word, for example, the word `potato`. Then, we will obtain the embedding of its mistyped version, for example, the word `potatoo`. We will then use cosine similarity to compute how semantically similar these two words are. Our assumption is that a good embedding model will be able to recognize that the second word is very similar to the first. A cosine similarity larger than `0.8` can reasonably be considered indicative of similar embeddings:

```
from sklearn.metrics.pairwise import cosine_similarity as sim

s1 = Sentence("eating potato")
s2 = Sentence("eating potatoo")

embedding = FlairEmbeddings('news-forward')
embedding.embed(s1)
embedding.embed(s2)
e1 = s1[1].embedding.tolist()
e2 = s2[1].embedding.tolist()

print(sim([e1], [e2]))
```

The preceding code prints out the cosine similarity between the word `potato` in the `eating potato` context and the word `potatoo` in the `eating potatoo` context. The similarity score printed out is `0.865`, indicating a strong connection between the two word embeddings.

Pooled Flair embeddings

For improved performance, you may also use `PooledFlairEmbeddings`. This uses the same exact syntax interface and supports the same set of models as `FlairEmbeddings`, but generally provides a slight improvement in performance. Do note that `PooledFlairEmbeddings` requires substantially more memory, meaning the marginal improvement in performance may not always be worth the extra memory consumption.

Available Flair embeddings

The set of supported Flair embeddings in the framework is ever-changing and a complete list would likely already be outdated by the time you read this book. But in general, you are likely to find the embedding you are looking for in one of the following categories:

- English Flair embeddings trained on a 1 billion-word corpus:

  - ID: `news-DIRECTION`

- Flair embeddings trained on Wikipedia, Opus, or other large corpora:

  - ID: `LANGUAGE_CODE-DIRECTION`.

    For example, for the forward version of German Flair embeddings, use `de-forward`.

- Flair embeddings trained on biomedical PubMed abstracts:

  - ID: `pubmed-DIRECTION`

- Multilingual Flair embeddings supporting 300+ languages:

  - ID: `multi-DIRECTION`

Note that the term `DIRECTION` needs to be replaced with either `backward` or `forward` when using the embedding ID.

For a complete and up-to-date list of supported Flair embeddings, you should check Flair's official documentation at `https://github.com/flairNLP/flair/blob/master/resources/docs/embeddings/FLAIR_EMBEDDINGS.md`.

In the preceding exercises, we covered the basic use of Flair embeddings. But you might have asked yourself: how can I use Flair embeddings in a way that uses the complete context appearing both before and after the word? This, and many more questions, will be answered in the next section.

Stacked embeddings

Stacked embeddings are a type of meta embeddings that allow us to form new embeddings by combining two or more embeddings together. These meta embeddings, as the name suggests, simply stack multiple embeddings on top of each other. They are ideal for combining the forward and backward versions of Flair embeddings. They can also be used for adding other embedding types to the mix, for example, to mix contextual string embeddings with the classic word embeddings. In fact, this is the type of a combination suggested by Flair and often the combination yielding state-of-the-art results.

To use stacked embeddings, simply instantiate a `StackedEmbeddings` class, passing in a list of embedding objects. You may then use this meta embedding in the same exact way you would use any other embedding in Flair.

Let's try this out on the set of embeddings recommended by Flair for best performance. We will stack `glove` word embeddings combined with `news-forward` and `news-backward` Flair embeddings:

```
from flair.embeddings import FlairEmbeddings, WordEmbeddings
from flair.embeddings import StackedEmbeddings


glove = WordEmbeddings('glove')
news_fw = FlairEmbeddings('news-forward')
news_bw = FlairEmbeddings('news-backward')


combined_embeddings_list = [glove, news_fw, news_bw]


stack = StackedEmbeddings(combined_embeddings_list)
```

You may then run `stack.embed(sentence)` on a `Sentence` object as you would using any other embedding types in Flair.

We have now covered the main word embedding types in Flair. It's now time to move one level up and learn how to compute embeddings not just for words but entire documents.

Document embeddings

Document embeddings are similar to word embeddings, with the only difference being that instead of getting one embedding for each word, we get one embedding for the entire document.

In Flair, a document can be defined inside the `Sentence` object. Instead of accessing the embedding through `sentence[n].embedding`, which would return the nth word's embedding, we simply run `sentence.embedding` or `sentence.get_embedding()`.

Flair currently supports the following embedding types:

- `TransformerDocumentEmbeddings`, a document embedding class using a pre-trained Hugging Face transformers.

- `DocumentPoolEmbeddings`, a meta document embedding class that takes a word embedding object, computes the embedding for each word, and returns the mean of all word embeddings.

- `DocumentRNNEmbeddings`, a meta document embedding class that takes a word embedding object, trains a recurrent neural network on the entire document, and returns the final state as the embedding for that document.

- `SentenceTransformerDocumentEmbeddings`, a document embedding class using the external `sentence-transformers` library. The library needs to be installed separately.

Flair recommends using the `TransformerDocumentEmbeddings` class for document classification tasks. Let's try it out:

```
from flair.data import Sentence
from flair.embeddings import TransformerDocumentEmbeddings

embedding = TransformerDocumentEmbeddings('bert-base-uncased')

sentence = Sentence('Example sentence .')
embedding.embed(sentence)

print(sentence.embedding)
```

The preceding code will print out a single embedding representing our sentence.

Note that in the preceding code example, we passed the `bert-base-uncased` embedding ID to the `TransformerDocumentEmbeddings` class. This is the name of the transformer model we decided to use. You may choose any other compatible model from the Hugging Face transformer models library.

Other embeddings in Flair

In this chapter, we covered most of the embedding types and techniques responsible for Flair's state-of-the art performance, though there are still many other embedding types left unexplained in this book.

These include the following:

- `ELMoEmbeddings`: Contextualized embeddings using a bidirectional LSTM

- `OneHotEmbeddings`: Embeddings where each word is a one-hot vector

- `BytePairEmbeddings`: Embeddings precomputed on the subword-level

- `TransformerWordEmbeddings`: Embeddings using transformer-based architectures

- `CharacterEmbeddings`: Character-level embeddings that are randomly set each time you initialize the class

While the inner workings of these embeddings exceed the scope of this book, even without fully understanding how they work, you should now be able to use them the same way you use any other embedding object in Flair as they all share the same interface.

Summary

In this chapter, we covered embeddings in Flair. They are the core concept behind all NLP tasks. We learned about how embeddings behave, their role, and how to use them in Flair through hands-on practical exercises.

We implemented an NLP analogy solver that can predict the last word in an analogy using arithmetic operations and relationships between word embeddings. By doing this, we not only refreshed our memory on how to use Flair's base types in practice, but we also illustrated what word embeddings are and how to derive meaning from them. We then also covered all the embedding types in Flair. We first covered classic word embeddings, followed by Flair embeddings. Using a meta embedding method, we used stacked embeddings and combined multiple embeddings into a new embedding – a technique used for achieving state-of-the-art results in Flair. We then gained hands-on experience of how to form document embeddings in Flair and concluded the chapter with a list of embedding types that exceed the scope of this book.

In the following chapter, we will learn about how Flair takes advantage of these embeddings to produce some of the best-performing sequence taggers in NLP.

4
Sequence Tagging

Sequence tagging (or **sequence labeling**) refers to a set of **Natural Language Processing** (**NLP**) tasks that assign labels or tags to tokens or other units of text. When the tags are named entities, we are then dealing with **named entity recognition** (**NER**). When the tags are parts of speech, this task is called **part-of-speech** (**PoS**) tagging. Unlike embeddings that are trained in an unsupervised manner, sequence taggers are trained using supervised training techniques, making them easier to evaluate and compare.

Sequence tagging is a field where **Flair** truly shines. Flair uses the ingenuity of Flair embeddings (explained in the previous chapter) to achieve state-of-the-art results across many different sequence tagging tasks and languages.

In this chapter, we are going to briefly explain how sequence taggers work in Flair. This will allow us to fully understand the inner workings as we cover NER, PoS tagging, chunking, and other sequence tagging techniques found in Flair. The knowledge gained in these sections will help you understand the range of sequence tagging tasks solved by Flair and what pretrained models are available for you to make use of. We will then show what metrics are generally used in NLP to measure and compare the performance of sequence taggers.

We will cover sequence tagging in Flair in the following sections:

- Understanding how sequence tagging works in Flair
- Named entity recognition in Flair
- Part-of-speech tagging in Flair
- Other sequence taggers in Flair
- Sequence labeling metrics

Technical requirements

For this chapter, you will require a **Python** environment with Flair version 0.11.

Code examples covered in this chapter are found in this book's official GitHub repository in the following **Jupyter Notebook**:

```
https://github.com/PacktPublishing/Natural-Language-
Processing-with-Flair/tree/main/Chapter04.
```

Understanding sequence tagging in Flair

Before diving straight into how to use sequence taggers in Flair, let's first briefly explain how sequence tagging (also known as sequence labeling) works. From an engineering point of view, a sequence tagger is a tool that receives text as input and returns a list of assignments, where each assignment includes a tag name and a span indicating where the annotated unit of text (usually a word) begins and ends.

The architecture and design of sequence labeling in Flair can best be explained with the following diagram:

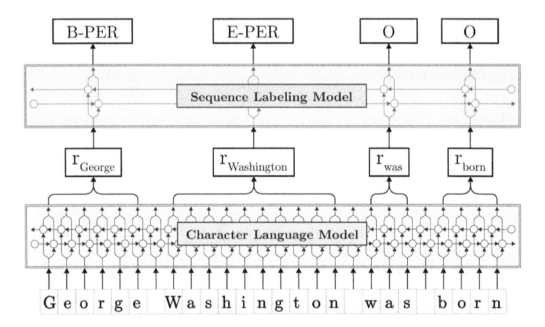

Figure 4.1– Overview of sequence tagging in Flair

The preceding diagram displays the sequence tagging architecture consisting of two components: the language model (in yellow) and the sequence tagging (labeling) model (in blue). The bidirectional character language model receives the original text as input and computes contextual embeddings (r_{George}, $r_{Washington}$, r_{was}, r_{born}) for each word. These are then passed into a **BiLSTM-CRF** sequence labeler. This special type of labeler is a novel sequence labeling approach that makes use of a **bidirectional LSTM (BiLSTM)** recurrent neural model combined with a **Conditional Random Field** (CRF) – a type of combinatorial algorithm. It outputs the probability distribution of labels corresponding to individual words. The most probable label is then chosen for each word (as seen in the preceding figure in the top row). For example, the word **Washington** is labeled as **E-PER** in the preceding figure, where **PER** suggests the word relates to a person and **E** marks the ending of a two-word named entity chunk. Words are often also tagged with a special label, **O** (meaning *other*), used for words not considered named entities or named entities belonging to none of the defined categories.

The preceding diagram illustrates the design of sequence taggers with one of the most common sequence tagging applications – NER. Let's learn how to use it in Flair in the following section.

Named entity recognition in Flair

NER is a special sequence tagging task that identifies and labels named entities. Named entities are real-word objects such as persons, locations, or organizations. The set of possible distinct named entities isn't strictly defined and may vary from model to model. Flair ships with a rich set of pretrained NER models, as well as tools to train custom models. In this section, we are going to focus on covering the pretrained models, as well as how to interpret and understand their output.

To get a good understanding of how NER tagging works in Flair, let's try it out as part of an exercise. We will start with tagging a single word that is clearly a named entity.

```python
from flair.data import Sentence
from flair.models import SequenceTagger


tagger = SequenceTagger.load('ner')
sentence = Sentence('Berlin')
tagger.predict(sentence)


print(sentence.to_tagged_string())
```

In the preceding script, we loaded Flair's default English NER tagger into memory by calling `SequenceTagger.load('ner')`. We then tagged a sentence with a single word – `Berlin`. Finally, we printed out the tagged sentence. The output is as follows:

```
Sentence: "Berlin" → ["Berlin"/LOC]
```

When using the `ner` tagger for the first time, you will also see a download progress bar and a notification that the model is hosted on the **HuggingFace ModelHub**.

The preceding NER tagging code snippet printed out: `"Berlin"/LOC`. This indicates that the tagger thinks the word `Berlin` in its given context, or lack thereof, is a location as it was tagged with the label `LOC`.

The set of **NER** classes varies from model to model, even within Flair, but most NER taggers classify entities with the following categories:

- Person
- Location
- Organization
- Miscellaneous

Let's have a look at another example, one that was used in the paper that introduced Flair to the NLP world. It presents a case where Flair's excellent word sense disambiguation performance truly shines:

```
tagger = SequenceTagger.load('ner')
text = 'George Washington was born in Washington.'
sentence = Sentence(text)
tagger.predict(sentence)

print(sentence.to_tagged_string())
```

The preceding script will print:

```
Sentence: "George Washington was born in Washington ." →
["George Washington"/PER, "Washington"/LOC]
```

The above output clearly shows how Flair was able to distinguish between `Washington` in the sense of a location (where it was tagged with `LOC`) and `Washington` in the sense of a person (where it was tagged as `PER`).

Types of NER taggers in Flair

Flair's pretrained NER models are classified as either monolingual or multilingual NER models.

Monolingual NER models are trained on a single language. They are tuned to achieve optimal performance for that single language while providing no support for any other language.

The most notable pretrained monolingual NER taggers are the following:

- English (4-class) NER tagger trained on the `Conll-03` dataset.

  ID: `ner`.

- English (4-class) NER tagger trained on the `Conll-03` dataset with reduced size, improved speed, but reduced accuracy.

  ID: `ner-fast`.

- English (4-class) NER tagger trained on the `Conll-03` dataset with improved accuracy, but reduced speed and a size increase.

  ID: `ner-large`.

- English (18-class) NER tagger trained on the `Ontonotes` dataset.

  ID: `ner-ontonotes`.

- German (4-class) NER tagger trained on the `Conll-03` dataset.

  ID: `de-ner`.

- French (4-class) NER tagger trained on the `WikiNER` dataset.

  ID: `fr-ner`.

- Spanish (4-class) NER tagger trained on the `Conll-03` dataset.

  ID: `es-ner-large`.

Unlike the preceding monolingual models, Flair also supports multilingual NER models. They do achieve a noticeably lower accuracy score for each specific language but come with an added benefit of allowing us to use one single model for a number of languages. Flair's multilingual NER models are trained over English, German, Dutch, and Spanish. The models can be loaded using the `ner-multi` and `ner-multi-fast` IDs.

You should now have a good understanding of how to use and interpret the results of NER taggers in Flair. We can now move on to taggers that, instead of tagging named entities, tag parts of speech.

Part-of-speech tagging in Flair

PoS taggers are responsible for tagging/labeling tokens with their corresponding parts of speech. A simplified set of parts-of-speech categories consists of nouns, verbs, adjectives, and adverbs. However, most taggers in Flair use larger, more descriptive tag sets. For example, the **Penn tag set** consists of 36 different tags.

The tagging process and PoS tagging syntax is no different from NER or any other tagging task in Flair. The only difference in loading a PoS tagger is choosing the right tagger ID that corresponds to our PoS tagger of choice.

Let's experiment with the default English PoS tagger in Flair using the pos tagger ID:

```python
from flair.data import Sentence
from flair.models import SequenceTagger


tagger = SequenceTagger.load('pos')
sentence = Sentence('Making a living')
tagger.predict(sentence)


print(sentence.to_tagged_string())
```

In the preceding script, we loaded Flair's default English pos tagger and tagged an example sentence. We then printed out the resulting tagged sentence. The output, along with the download bar when using the tagger for the first time, is as follows:

```
Sentence: "Making a living" → ["Making"/VBG, "a"/DT, "living"/
NN]
```

From the preceding output, we can see that our sentence, consisting of the Making, a, and living tokens, was tagged with the VBG, DT, and NN tags respectively. This means that the word Making was tagged as a verb in the present participle, the word a as a determiner, and the word living as a noun in the singular. This shows that the tagger handles word sense disambiguation well as the word *living* can either be a verb or a noun depending on the context.

To fully understand the meaning of each individual tag produced by English PoS taggers in Flair, we need to refer to the documentation of the Penn tag set.

Part-of-speech tagging tag sets

The PoS tagger in the preceding script uses the **Penn Treebank Project PoS tag set**. It is a standardized set of tags consisting of 36 different PoS tags. It allows researchers to use a standard that enables them to compare results across different taggers. It was originally used to tag the corpora in **The Penn Treebank Project**, but is today used across many different datasets.

The Penn tag set consists of the following 36 tags:

- CC: Coordinating conjunction
- CD: Cardinal number
- DT: Determiner
- EX: Existential there
- FW: Foreign word
- IN: Preposition or subordinating conjunction
- JJ: Adjective
- JJR: Adjective, comparative
- JJS: Adjective, superlative
- LS: List item marker
- MD: Modal
- NN: Noun, singular or mass
- NNS: Noun, plural
- NNP: Proper noun, singular
- NNPS: Proper noun, plural
- PDT: Predeterminer
- POS: Possessive ending
- PRP: Personal pronoun
- PRP$: Possessive pronoun
- RB: Adverb
- RBR: Adverb, comparative
- RBS: Adverb, superlative

- RP: Particle
- SYM: Symbol
- TO: To
- UH: Interjection
- VB: Verb, base form
- VBD: Verb, past tense
- VBG: Verb, gerund or present participle
- VBN: Verb, past participle
- VBP: Verb, non-third-person singular present
- VBZ: Verb, third-person singular present
- WDT: Interrogative word determiner
- WP: Interrogative word pronoun
- WP$: Possessive interrogative word pronoun
- WRB: Interrogative word adverb

Most English PoS taggers are likely to use this or a slightly modified version of this tag set. However, unlike NER tagging where the same NER classes are often used across several languages, the Penn set tags are very English-specific and cannot be effectively directly used on many non-English languages. A solution to this problem is the Universal PoS tag set. It defines a reduced universal tag set that can be used across multiple languages.

The Universal PoS tag set defines the following PoS tags:

- VERB: Verbs (all tenses and modes)
- NOUN: Nouns (common and proper)
- PRON: Pronouns
- ADJ: Adjectives
- ADV: Adverbs
- ADP: Adpositions (prepositions and postpositions)
- CONJ: Conjunctions
- DET: Determiners
- NUM: Cardinal numbers

- PRT: Particles or other function words
- X: Other: foreign words, typos, abbreviations
- .: Punctuation

The preceding tags are less English-specific and can be used across multiple languages. They are used for Flair's multilingual PoS taggers (described in more detail in the following section).

Types of PoS taggers in Flair

Flair ships with a number of pretrained PoS taggers. Just like NER taggers, Flair's PoS taggers can be classified as either monolingual or multilingual taggers.

Monolingual pretrained PoS taggers in Flair are trained on a single language and often use language-specific tag sets.

Some of the commonly used taggers are described as follows:

- English PoS tagger trained on the Ontonotes dataset using the Penn tag set.

  ID: pos

- A faster, lighter, and less accurate version of the English PoS tagger trained on the Ontonotes dataset using the Penn tag set.

  ID: pos-fast

- English PoS tagger trained on using the Universal POS tag set.

  ID: upos

- A faster, lighter, and less accurate version of the upos tagger using the Universal POS tag set.

  ID: upos-fast

- German PoS tagger trained on the UD German – HDT dataset.

  ID: de-pos

The preceding taggers will achieve optimal performance on the language they were trained on, but will not work well on any other language. To deal with multilingual corpora, Flair offers a set of multilingual PoS taggers. They were trained on 12 languages: English, German, French, Italian, Dutch, Polish, Spanish, Swedish, Danish, Norwegian, Finnish, and Czech. To load them, simply use the pos-multi or pos-multi-fast tagger IDs.

Other tagging tasks in Flair

While PoS and NER tagging are the two most commonly used sequence tagging tasks, Flair supports many others. This is possible in Flair because its sequence tagging architecture isn't tailored to a specific sequence tagging task. It can work with any, as long as the training corpus and annotations are stored in a format compatible with Flair.

Other sequence tagging tasks solved by Flair's pre-trained models include chunking, semantic frame detection, and domain-specific taggers such as negation and speculation tagging in biomedical texts.

Each sequence tagger mentioned in Flair's official documentation is accompanied by an accuracy score. Depending on what sequence tagger you are interested in, the metrics used to compute it might be different. In order to fully understand what the scores mean, we need to understand the metrics used to compute them. We will be covering that in the following section.

Sequence labeling metrics

When comparing one sequence tagger to another, we can't simply try them out by hand and take a wild guess about which one performs better. Their performance needs to be evaluated using the same dataset and computed using the same predefined metric. The most common metrics used in sequence labeling are *accuracy* and *F1 score*.

Measuring accuracy for sequence labeling tasks

Accuracy is a measure ranging from 0 to 1 that simply computes the proportion of correctly tagged tokens.

Assuming `correctly_tagged_tokens` is the number of correctly tagged tokens and `all_tokens` is the total number of all tokens, accuracy can be defined as:

$$accuracy = \frac{correctly\_tagged\_tokens}{all\_tokens}$$

The preceding formula is simple and provides an easily interpretable result, but the metric can be misleading when dealing with imbalanced datasets (datasets where classes/tag names are not represented equally). This is particularly noticeable with NER where the majority of tokens belongs to a single class. For example, a NER tagger with 85% accuracy sounds like a pretty good tagger. But when we look at the distribution of tags in a NER dataset, we will notice most of the tokens are tagged with the **O** (**other**) tag representing non-named entities. Assuming there are 85% of non-named entities in our corpus, a model that tags all tokens with **O** will achieve 85% accuracy by default. To solve this problem, we can use an alternative metric called the *macro F1 score*.

Measuring the F1 score for sequence labeling tasks

F1 score is the harmonic mean of two other metrics called *precision* and *recall*. It ranges from 0 (worst possible score) to 1 (best possible score).

In sequence tagging, *precision* is defined as the percentage of named entities found by the learning system that are correct. It is defined as the following:

$$precision = \frac{tp}{tp + fp}$$

Recall is the percentage of named entities present in the corpus that are identified by the system. It is defined as:

$$recall = \frac{tp}{tp + fn}$$

The following apply:

- **True positives** (*tp*): The number of tokens that are correctly labeled with a specific tag. For example, if a token is labeled as noun in the training data and our tagger labels it with the noun tag, this is considered a true positive.

- **False positives** (*fp*): The number of tokens that were incorrectly labeled with a specific tag. For example, if a token is labeled as verb in the training data, but our tagger labels it with the noun tag.

- **False negatives** (*fn*): The number of tokens that should have been labeled with a specific tag, but weren't. For example, if a token is labeled as noun in the training data, but our tagger labels it with the verb tag.

F1 score is then defined as the harmonic mean of precision and recall.

$$F1 = 2 \cdot \frac{precision \cdot recall}{precision + recall}$$

The process of working out the F1 score therefore involves the following:

1. Counting true positives, false positives, and false negatives
2. Getting *precision* and *recall* using the above formulae
3. Computing the F1 score as the harmonic mean of *precision* and *recall*

If we compute the F1 score for each tag (for example, for each named entity) separately, and then average all the F1 scores, we obtain the macro F1 score. This metric takes into account unequal representations of tags in the corpus.

If we compute the F1 score for all the tags counting the total true positives, false negatives, and false positives together, we obtain the *micro F1 score*. This metric will have the same value for accuracy as single-label problems (for taggers where one token can have at most one tag assigned).

Next time you assess reported sequence tagger results, be careful about whether the F1 score refers to the micro or the macro variant. The variant most often used in Flair's results (when not explicitly specified) is the micro F1 score.

Summary

With the knowledge gained in this chapter, we are not only able to use many of the pretrained sequence taggers available in Flair, but we are also capable of interpreting and understanding their output.

In this chapter, we covered sequence tagging in Flair from different angles. We touched on the design and architecture of Flair's proposed sequence taggers. We then covered NER motivation and theory, and what pretrained NER taggers can be found in Flair. We did the same with PoS tagging, where we covered the most notable tag sets, such as the Penn tag set and the universal PoS tag set, before covering the important pretrained PoS taggers found in Flair. We then finally studied the two metrics most often used to evaluate sequence taggers, and we emphasized the importance of distinguishing between micro and macro F1 scores.

In the next chapter, we will be looking into using Flair to train, store, and reuse your own sequence labeling models.

Part 2: Deep Dive into Flair – Training Custom Models

In this part, you will learn how to train custom sequence labeling models, embeddings, and classifiers with Flair, evaluate their performance, and achieve optimal results with hyperparameter optimization.

This part comprises the following chapters:

5
Training Sequence Labeling Models

In this book, we have so far covered a number of features and models Flair offers right out of the box. The set of pre-trained sequence taggers, embeddings, and other models at first seems large enough for us to never need anything more than what's already available in Flair. But chances are, if you work with **natural language processing** (**NLP**) for long enough, you will encounter a problem that isn't generic enough to be solved by pre-trained models. When faced with such a problem, we usually need to train our own. This process involves acquiring training data, preprocessing it, possibly hand-labeling it, and finally, working with an NLP framework to train the model. But because Flair ships a wide selection of labeled corpora, chances are that you will only ever need to perform the last step, model training – the main focus of this chapter.

We will start this chapter by explaining when and why training custom models is necessary. Next, we will go through the hardware requirements for training sequence labeling models, showing both the free and paid approaches to training your models in the cloud.

We will then briefly cover the process of neural model training. After that, we will look at the metrics and parameters that govern training. We will go through the actual model training syntax as part of a hands-on exercise where we train an English part-of-speech tagger. Finally, we will learn a few basic approaches to assessing model quality and running visualizations that will help us decide whether our model is good enough or whether we should continue training or start from scratch.

In this chapter, we will cover the preceding topics as part of the following sections:

- The motivation behind training custom models
- Understanding the hardware requirements for training models
- Working with parameters for training and evaluation
- Training custom sequence labeling models
- Knowing when to stop and try again

Technical requirements

All of the Python libraries used in this chapter are direct dependencies of Flair 0.11 and require no special setup, assuming Flair is already installed on your machine. Code examples covered in this chapter are found in this book's official GitHub repository in the following Jupyter notebook: `https://github.com/PacktPublishing/Natural-Language-Processing-with-Flair/tree/main/Chapter05`.

Running the code covered in this chapter on a GPU-enabled machine is not mandatory, though it is advised, as it will speed up the training process and allow you to train higher-quality models.

The motivation behind training custom models

In this book we so far discussed the features and models that are available in Flair straight out of the box. However, if you work with NLP long enough, you will likely encounter a sequence labeling problem that is complex or specific enough that there will be no pre-trained models available out there. This can happen in either of the following situations:

- *The problem you are solving is domain-specific*: Sequence taggers such as **Named Entity Recognition** (**NER**) or **Part-of-Speech** (**PoS**) taggers are usually trained on large, generic corpora that are supposed to represent the general use of a language. But if our problem is domain-specific, it's likely that we will require a custom tagger with domain-specific labels trained on domain-specific corpora.

- *A pre-trained model exists but doesn't perform well enough*: Every model made available in Flair and the approaches used for training are usually reviewed by other contributors, but it's entirely possible you may be able to build a better model yourself. Chances for this increase as you move toward the less popular languages, since their performance will not be as optimized. It may also happen that the pre-trained model achieves better overall performance in terms of the F1-score or accuracy but may fall short in tagging that one specific label that you really care about.

- *The language you're working on is less widely spoken and isn't supported by the framework*: When it comes to advances in NLP, the new, state-of-the-art solutions are often built for English or popular western languages. It is only when a technique gains popularity and traction that the models start being built for the underrepresented languages as well. If you are working with a rare language, chances are, you will need to train your own tagger.

- *You want to contribute to Flair and the open source community*: Finally, you may not really have a need to build a specific tagger for a specific language. You may simply wish to contribute to Flair. Given that there's a vast number of languages not supported by Flair's pre-trained models yet, you can just pick an underrepresented language and get your hands dirty.

Whatever your motivation to train custom sequence taggers is, the process of getting there will be the same. First, let's make sure we have the right tools for model training by getting an understanding of the hardware requirements in the next section.

Understanding the hardware requirements for training models

Flair sequence labeling models are essentially special types of neural networks. You may have heard that in order to do inference on (that is, use) or train neural networks, you need a high-performance **Graphical Processing Unit** (**GPU**)-equipped machine. Training a neural network requires a computation of a large number of mathematical operations (largely matrix multiplication). Most of these operations can be parallelized much better on GPUs as opposed to CPUs, which speeds up the training process significantly. But this doesn't necessarily mean you can't do any training or inference without a GPU. Whether you actually need a GPU will simply depend on the size of the neural network you are training and the number of hours, days (or decades) you have at your disposal to wait for the training to finish. If you are simply starting off with neural networks and are experimenting with training tiny networks with a handful of layers and a few nodes, you should be completely fine with even a slightly older PC without a dedicated GPU.

Unfortunately, most models in Flair will be too large to train on a CPU in a reasonable amount of time. Doing inference on a CPU is perfectly doable and single samples can be tagged within seconds, but training on a CPU with the goal of producing a good model can take weeks, if not months. In fact, even an average GPU can take from a day to several days to train a sequence labeling model.

It is now clear that training Flair models that yield good results will require a GPU-equipped machine. But buying a GPU and sticking it into a PC is not your only option. You can choose from either of these:

- *Free Jupyter notebook environments such as Google Colaboratory*: **Google Colaboratory** (also known as **Colab**) is a free Jupyter notebook environment where you can run GPU-equipped Python sessions. The environment also offers a decent amount of RAM and a good chunk of CPU power. But it does come with its limitations. You can only run sessions of up to 12 hours. This means that if you train a model for over 12 hours and fail to save it before that, your work will be lost. Luckily, Flair offers a special solution called **checkpoints** that cause the model trainer to store a model snapshot after each epoch. Training can then be resumed from the last successful epoch, even if your session expired after 12 hours. We recommend connecting your Colab sessions to Google Drive so that your models are automatically saved to the cloud.

- *Free cloud computing credits*: You will be awarded a certain number of credits when you sign up for the first time with cloud service providers such as Microsoft Azure. Azure allows you to use those credits to spin up GPU-equipped **Virtual Machines** (**VMs**) on which you can easily run Flair. The downside is that you will eventually run out of free credits, which will leave you with the next option.

- *Paid-for cloud services*: There is a large number of paid-for cloud services that offer either pre-set Jupyter notebook environments or the very basic GPU-equipped VMs. The advantage of the paid-for cloud services is that your sessions are not time-limited. Another great perk is also that if you find out that the environment you are using doesn't have enough resources, you can simply swap it with a more powerful machine.

- *Building your own deep-learning machine*. By far the biggest perk of building your own rig is that once you pay the initial cost of hardware, you will be able to train as many neural networks for as long as your heart desires. The downside of this approach is that your hardware will eventually become outdated, and you will soon find yourself in a situation where your machine simply isn't powerful enough anymore. The only solution then will be to upgrade by buying new hardware.

- *Training models on CPU for learning and experimentation.* This last option is a possibility for anyone who would simply like to get a taste of model training. Because model training on CPUs is slow, you will need to train on smaller corpora for fewer epochs. One technique that allows that is downsampling. When you downsample a dataset (for example, by calling `dataset.downsample(0.001)`), you end up with only a fraction of the original dataset which may be small enough even for a CPU. This does mean that the models you produce are unlikely to ever achieve good results. But you will still get a glimpse of what model training looks like and will have the know-how and knowledge to do it for real once you do get access to a more powerful setup.

While some of the preceding options will allow you to train state-of-the-art models and some will merely suffice for learning and experimentation, all are sufficient for following the upcoming hands-on section on how to train sequence labeling models in Flair. But before we do that, we need some understanding of the basic math, metrics, parameters, and deep-learning jargon.

Working with parameters for training and evaluation

Flair is arguably one of the simplest NLP frameworks out there. But to produce good performing taggers, we still need to have some level of understanding of the underlying concepts and parameters passed to Flair.

First, let's quickly explain what model training really is.

Understanding neural model training

Neural networks are a special subset of machine learning. They are loosely based on biological neural networks such as the human brain. Neural nets generally comprise the network architecture and their weights and biases.

The network architecture is all that defines the design of the network. It includes everything from the number of layers and the number of input and output nodes to the types of units used. These properties are referred to as hyperparameters. They remain unchanged for a single model training session.

Then, there are weights and biases – they are the final result of model training. The process of model training is essentially a process of computing the optimal weights. Training starts by first initializing all the weights and biases with random values. We then repeat the following steps a number of times:

1. The model is fed a batch of training data samples for which we can compute output. In the case of sequence labeling models, the output is token labels.

2. The output (predicted labels) is then compared to the target values (actual labels) and evaluated using the loss function also known as the `objective` function. It tells us how close our predictions are to the ground truth.

    The goal of the model trainer is to find weights that result in the smallest loss possible. A loss of zero would be an ideal loss. It would mean all predictions align with the target values in the training data.

3. Once loss for a single batch of training is computed, the network tweaks the weights in the hope of improving the model, using a technique called backpropagation.

4. We then feed the next batch of training data to the model and repeat the process from *step 1* onwards, until we run out of all training data.

The preceding sequence of steps in which all training data is consumed for training is called an **epoch**. But because the changes applied to the model weights are very small, one pass through the data usually isn't enough. We usually keep repeating the training process for many more epochs until the loss function converges and stops improving. The optimal number of epochs will largely depend on the type and amount of data, the number of tags, and the hyperparameters used. At the end of each epoch, we also evaluate the model on the test dataset using a performance metric such as accuracy or the F1-score.

When training completes, the weights stored in memory are then usually written to a file so that they can be loaded when we want to do inference on the model or simply continue training.

The process described in this section may seem complex and overwhelming. But luckily, all of it is implemented and automated as part of the framework. So, not having a complete understanding of the underlying logic behind neural network training is not a deal-breaker for anyone hoping to build sequence taggers. What is important is understanding the purpose and meaning of the parameters and values passed to the model training classes and methods in Flair.

Let's cover these parameters one by one.

Parameters and terminology used in Flair model training

Here are some common hyperparameters, parameters, and general machine learning concepts you will encounter as part of training models in Flair:

- **Learning rate**: It is possibly the most important parameter in model training. It's essential to get this value right. If the learning rate is set too low, the model will require a huge number of epochs, and the tiny increments will lead to a slow learning progress. A low learning rate, on the other hand, can cause our model to get stuck in a local minimum (a seemingly best solution that can likely be improved by exploring more options). If the learning rate is too high, it will cause the optimizer to adjust the weights too aggressively, causing the corrections to the weights to overshoot. It will cause the loss to fluctuate all over the place across different epochs and fail to converge to a minimum. Choosing the right learning rate is therefore essential.

- **Mac epochs**: This is a setting that determines how many passes through the training data the trainer will make before finishing. This setting affects training time linearly. When set too high, training can take longer than we were prepared to wait for. When set too low, the trainer will finish before it had a chance to find the optimal weights.

- **Loss**: This is a measure of how well our model is doing compared to the ground truth. The value is usually averaged across several samples. The goal of model training is to minimize loss. If the loss value is zero, it would mean that our model predicted all the tags correctly. Loss is often plotted on a graph to illustrate the progress and trend of learning. Seeing a decrease in loss value usually indicates that the model is learning, but it can also mean that it is merely overfitting our training data.

- **Mini batch size**: When training models, we are unlikely to feed the entire training set to the model at once. Instead, we may, due to memory limitations, choose to split the training set into smaller batches. We can then feed each batch of data to the model individually. The size determining the number of learning samples in a single batch is called the **mini-batch size**.

- **Patience**: This is the number of bad epochs (epochs without improvement to the objective function) to train before stopping.

- **Hidden size**: In deep learning, this refers to the number of hidden layers or number of hidden **Recurrent Neural Network** (**RNN**) states. This setting is supplied as a parameter to the tagger classes as well as some embedding classes, such as **FlairEmbeddings**.

- **Checkpoints**: These are temporary snapshots of the model usually done at the end of an epoch. Checkpoints can be useful if we think our training session may get interrupted, which will save us from having to start all over again. We can simply load the checkpoint at the beginning of the new training session and continue from where we left off before our session was interrupted.

The terminology explained in the preceding list covers the concepts found throughout many machine learning frameworks, including Flair. Having a rough understanding of their role in model training means that you're ready for the real thing. You are now ready to train sequence labeling models in Flair. Let's dive right in.

Training custom sequence labeling models

In this section, we will be looking at the process, syntax, objects, and methods involved in training custom sequence labeling models in Flair. If you read the previous sections of this chapter and understood the contents, you're in luck. Once you understand the underlying concepts of neural networks, have a GPU-equipped rig running, and are familiar with the most common parameters, the actual training process is actually fairly straightforward.

The process can be broken down into the following steps:

1. Loading a tagged corpus
2. Loading the tag dictionary
3. Building the embedding stack
4. Initializing the `SequenceTagger` object
5. Training the model

Each of these steps requires only a few lines of code. To best understand the code, let's cover it as part of a practical example of training a PoS tagger. For example, let's pretend there are no pre-trained English taggers in Flair and attempt to train a PoS tagger for English from scratch.

We will be following an adapted version of Flair's recommended training steps found in Flair's GitHub repository documentation.

Loading a tagged corpus

When attempting to load a tagged corpus in Flair, we need to do the following:

1. We first need to verify whether such a corpus even exists and exactly how is it named. This information can be found in the official documentation on Flair's GitHub page at `https://github.com/flairNLP/flair/blob/master/resources/docs/TUTORIAL_6_CORPUS.md`.

    For example, the default corpus for English is called `UD_ENGLISH`. It can be loaded with a single function call:

    ```
    from flair.data import Corpus
    from flair.datasets import UD_ENGLISH


    corpus = UD_ENGLISH()
    ```

    This will, if executed for the first time, download the model from internet storage, which may take a few minutes.

    The model will then be stored for future use, meaning that the next time you run the same command, it will load the model from disk and execute within seconds.

2. Because our corpus contains a wide range of different tags and labels, we need to decide what the goal of our tagger will be so that we can filter out all irrelevant tags. We will do that by storing the type of tagger used into a variable. This variable will later be passed to different methods:

    ```
    tag_type = 'pos'
    ```

    The `tag_type` variable set to `pos` indicates we are dealing with PoS tagging.

3. Finally, we need to decide whether we will be aiming for a state-of-the-art model or whether we simply want to give Flair a go in order to gain experience in training models without the expectation of good results. If you fall into the latter category, you can vastly reduce your training time by simply downsampling your dataset. We can, for example, reduce the dataset to 1% of its original size and reduce the training time from hours to minutes by running the following:

    ```
    import random


    random.seed(123)   # ensure we get same split every time
    corpus.downsample(0.01)
    ```

This will allow us to train the tagger even on a CPU. Note that in the preceding script, aside from downsampling the corpus, we also imported the `random` library and set a random generator seed. We downsampled our dataset so that we can go through this exercise quicker without having to wait for long periods of time for the models to train. When training neural models on small amounts of data, we don't always get the expected results, and the training process can be unstable. This means that if you choose a slightly different subset of the dataset, the results may be entirely different. To counter this issue, we used the `random.seed()` method, which ensures that Flair's randomized subset selection method behaves consistently every time.

Our corpus is now ready. Let's move on to the next data loading and preprocessing step.

Loading the tag dictionary

A tag dictionary is simply a set of all possible tags. For PoS, this includes categories such as verbs, adjectives, nouns, and interjections. It also includes a few special tags, such as the `unk` tag, indicating unknown words that we can't compute classic word embeddings for.

A `tag` dictionary can be acquired by running the following code block:

```
tag_dictionary = corpus.make_label_dictionary(tag_type)
print(tag_dictionary)
```

The preceding snippet will print out the following result:

```
Dictionary with 50 tags: <unk>, O, DT, NN, VBZ, RB, NNP, CD, .,
WP, PRP, JJ, '', '', IN, CC, NNS, ,, VBP, MD, VB, TO, POS, WRB,
VBD, VBG, NNPS, VBN, HYPH, -LRB-
```

The preceding output shows that we are dealing with 50 different PoS tags where a few (such as the `unk` tag) have special meaning and do not represent parts of speech.

We can now move on to the next section, in which you will greatly benefit from the knowledge gained from *Chapter 3, Embeddings in Flair*.

Building the embedding stack

The ability to build embedding stacks is one of the great features in Flair that often contributes to it achieving state-of-the-art results. A stack can be built by simply passing a list of Flair-supported embedding objects to a `StackedEmbeddings` class constructor.

In this example, we are going to build a stack of non-contextual English **FastText** embeddings (the default embeddings when using language codes such as en), contextual English forward Flair embeddings, and contextual English backward Flair embeddings. This can be done by running the following code block:

```
from flair.embeddings import (WordEmbeddings,
                              StackedEmbeddings,
                              FlairEmbeddings)

embedding_types = [
    WordEmbeddings('en'),
    FlairEmbeddings('news-forward'),
    FlairEmbeddings('news-backward'),
]

embeddings = StackedEmbeddings(embeddings=embedding_types)
```

Flair will, just like when loading corpora, download the web-hosted embedding objects and store them for future use. This step can take up to a few minutes.

We can now move to the last two preparation steps before the actual training.

Initializing the SequenceTagger object

We will use the same SequenceTagger class that we previously used to load pre-trained models in *Chapter 4*, *Sequence Tagging*, but this time, to define entirely new models. If you remember correctly, we used a static call to SequenceTagger.load() to create tagger objects from pre-trained models. Here, we will initialize the class directly by passing in the properties that define our model. These properties define everything from the architecture of the model, the types of recurrent units, and the number of hidden RNN states to the set of possible tags predicted by the model.

The required parameters of the SequenceTagger class are as follows:

- hidden_size: This defines the number of hidden states in an RNN. A common number to start with is 256, but the optimal value will completely depend on the problem you are solving.

- embeddings: The embeddings class needs to be an instance of the TokenEmbeddings class. It can either be CharacterEmbeddings, FlairEmbeddings, WordsEmbeddings, or the class we are using in our example – StackedEmbeddings.

- `tag_dictionary`: The dictionary containing all the possible tags that we generated in the previous section using the `corpus.make_label_dictionary()` method.

- `tag_type`: A string representing the type of tagger we are training. For PoS tagging, use `pos`, and for NER, make sure to use `ner`.

To initialize a `SequenceTagger` object for the new model we are training as part of this exercise, run the following code block:

```
from flair.models import SequenceTagger

tagger = SequenceTagger(hidden_size=256,
                        embeddings=embeddings,
                        tag_dictionary=tag_dictionary,
                        tag_type=tag_type)
```

Here, `embeddings`, `tag_dictionary`, and `tag_type` are the variables defined in the preceding bullet list.

And that's it! The preceding code should take no longer than a second to execute. We now created an English PoS sequence tagger, fully capable of tagging English text with PoS tags.

If the preceding paragraph made you raise an eyebrow, you're not alone. Sequence tagging models should take a not-insignificant amount of resources and time to train – no sequence labeling model should take less than a second to produce, right? Well… not quite. A *good* sequence labeling model will take a significant amount of resources and time to produce. But a *terrible* one doesn't have to! What we produced with this code is a complete and fully functioning sequence labeling model with randomly initialized weights. This means that the model is perfectly capable of tagging text; it's just that the accuracy of the results will be as good as a random guess.

In the following example, we will use our tagger with randomly initialized weights to tag some text:

```
from flair.data import Sentence

sentence = Sentence('Hello world')
tagger.predict(sentence)
print(sentence)
```

This code block will spit out something like the following:

```
Sentence: "Hello world" → ["Hello"/RB, "world"/VB]
```

Here, the words Hello and world are tagged with RB and VB, suggesting that the first word is an adverb and the second one is a verb. This is clearly incorrect. If you run these two code snippets yourself, you will likely get different results. This is because the model weights of our freshly generated model are randomly initialized. Unless you are extremely lucky, your results will also be incorrect.

To make our model produce good results, we need to tweak the model weights in such a way that they will produce as accurate results as possible. This is done through model training. Let's figure it out.

Training the model

We now have a tagged corpus that is loaded into memory, embeddings are defined, and the SequenceTagger object is created. We're now ready to start training.

The process of training involves first creating a ModelTrainer object and then calling its train() method.

ModelTrainer expects the following required parameters:

- model: A Flair model class – for example, a SequenceTagger object
- corpus: A Corpus object

Knowing all of this information, we can now initialize the ModelTrainer object by executing the following lines of code:

```
from flair.trainers import import ModelTrainer

trainer = ModelTrainer(tagger, corpus)
```

This leaves us at the last step of sequence labeling model training – calling the train() method. It receives a ridiculously long number of parameters, but here are some you might find useful:

- base_path: The path to where the model will be stored. If the directory, or the subdirectories within it don't exist, they will be created.
- learning_rate: The initial learning rate. It defaults to 0.1.

- `mini_batch_size`: The size of the mini-batches during training. It defaults to `32`.

- `max_epochs`: The number of epochs before finishing training. `150` is a good number to start with.

- `patience`: The number of epochs with no improvement before terminating. It defaults to `3`.

- `train_with_dev`: Whether to use both the dev and train datasets for training. The dev dataset (sometimes also called the validation dataset) is a dataset used for tuning hyperparameters. If you aren't doing any hyperparameter tuning, you can use that dataset for training instead. This allows you to train the model on a larger amount of data.

- `checkpoint`: If `True`, Flair will store a model snapshot named `checkpoint.pt` in `base_dir` at the end of each epoch so that training can be resumed in case it gets interrupted.

- `write_weights`: If `True`, Flair will write weights to a file called `weights.txt`.

- `optimizer`: A **PyTorch** optimizer class. It uses stochastic gradient descent by default.

- `epoch`: The starting epoch used if continuing a training run that was interrupted.

- `use_tensorboard`: Writes out **TensorBoard** information if `True`.

- `tensorboard_log_dir`: The directory where the TensorBoard data may be stored.

- `metrics_for_tensorboard`: Custom optional metrics for TensorBoard.

But What Is TensorBoard?

TensorBoard is a web tool for inspecting and understanding neural network training runs. It plots all the training data, such as loss, accuracy, and many other metrics, and allows us to easily monitor the training process. The tool was primarily built for integration with the **TensorFlow** deep learning framework but has since been integrated into many other tools. One of the tools that integrated TensorBoard is PyTorch – the underlying deep learning framework powering Flair. That's why hooking TensorBoard up with Flair is as easy as enabling it by setting the `Tensorboard` argument to `True` and passing the log directory of your running **Tensorboard** instance to the `ModelTrainer` constructor.

We can now finally start the actual training of our model:

```
trainer.train('tagger',
              learning_rate=0.1,
              mini_batch_size=32,
              max_epochs=150,
              train_with_dev=True)
```

This will start the training process and run it for, at most, 150 epochs. It will store the model results in the tagger directory, which will be created inside the directory you are running the training script from.

If everything is done correctly, the trainer should resume the first epoch. When the first two epochs are done, the output should look similar to this:

```
------------------------------------------------------------
epoch 1 - iter 1/4 - loss 65.13372040 - lr: 0.100000
epoch 1 - iter 2/4 - loss 69.70247269 - lr: 0.100000
epoch 1 - iter 3/4 - loss 71.33122508 - lr: 0.100000
epoch 1 - iter 4/4 - loss 69.54229164 - lr: 0.100000
------------------------------------------------------------
EPOCH 1 done: loss 69.5423 - lr 0.1000000
DEV : loss 47.70802688598633 - score 0.0654
BAD EPOCHS (no improvement): 0
------------------------------------------------------------
epoch 2 - iter 1/4 - loss 53.46533203 - lr: 0.100000
epoch 2 - iter 2/4 - loss 59.09352493 - lr: 0.100000
epoch 2 - iter 3/4 - loss 58.95171865 - lr: 0.100000
epoch 2 - iter 4/4 - loss 59.12093544 - lr: 0.100000
------------------------------------------------------------
EPOCH 2 done: loss 59.1209 - lr 0.1000000
DEV : loss 41.87521743774414 - score 0.1115
BAD EPOCHS (no improvement): 0
```

The first and probably most important thing to monitor during training is the trend of how a loss value is changing. A decrease in loss usually means the model is learning. From the previous output, we can see that the first epoch finished with a loss of `69.5423`, and the second epoch finished with a loss of `59.1209`. This indicates that the model is learning! The output also indicates that our accuracy score increased from `0.0654` to `0.1115`.

If you remember, earlier in this section, we mentioned that the sequence tagger objects are initialized with random weights. This means that when you run the same exact code twice, the values will be slightly different.

Training, unless interrupted, will now continue to pass through a number of epochs until it does `max_epochs`, or until the trainer fails to improve the model for `patience` consecutive epochs.

On an average GPU, training a downsampled English dataset should finish within a few minutes. When it does, Flair will report the final results of the best model tested on the test dataset. The results should look something like this:

```
Results:
- F-score (micro): 0.8878
- F-score (macro): 0.702
- Accuracy (incl. no class): 0.8878
```

After these results, we should also see results reported for each tag individually.

Training is now complete. We can now try to tag the same pair of words (`hello world`) `sentence` on our newly trained model to see whether the results are now any better:

```
tagger.predict(sentence)
print(sentence)
```

This code will, if you followed the steps correctly, print out the following:

```
Sentence: "Hello world" → ["Hello"/UH, "world"/NN]
```

The UH tag in the output stands for interjection and NN for noun. This means that our trained tagger labeled our example sentence correctly!

> **Important Note**
> In the unlikely case that you ran this code and your results happen to be different, this is simply because you ran your tagger on too small of a dataset. To achieve the same results as the preceding output, make sure to restart the training process on a dataset that is downsampled to a lesser extent.

In the preceding code snippet, we tagged text by referring to the `tagger` variable, which holds the model stored in memory. But on top of storing our trained model in memory, Flair also stored it in the filesystem so that it can be loaded and used at a later point.

Understanding training output files

After a successful training session, Flair will have written a number of training output files to a directory defined by the `base_dir` parameter. They, aside from the actual model, hold important information about training.

Here is a list of files to expect in `base_dir` after a training run:

- `final-model.pt`: The actual model. This can also be called `best-model.pt` when using dev data for hyperparameter tuning.

- `loss.tsv`: A tab-separated document containing information about the training loss for each epoch.

- `training.log`: A complete log of standard output written out during training.

- `test.tsv`: A tab-separated document containing tokens, their real tags, and their predicted tags. This document allows you to do token-level analysis, verifying which tokens were predicted correctly and which weren't.

- `weights.txt`: Model weights. This file will only be populated only if `write_weights` was set to `True` during training.

With our trained model now stored in the filesystem, we can now test loading the model from a file in a fresh Python session.

Loading and using custom Flair models

In the final stages of model training, Flair will write a file called `final-model.pt` to a path defined by the `base_dir` argument of our `ModelTrainer`. Loading this model into Flair is as easy as passing it as the first argument to the `SequenceTagger.load` method.

Assuming `tagger` is our `base_dir` and our Python session's current working directory is the same as the one we were in when doing training, the model can be loaded by running the following:

```
from flair.models import SequenceTagger

tagger = SequenceTagger.load('tagger/final-model.pt')
```

We can then test the tagger out by running the following:

```
from flair.data import Sentence

sentence = Sentence("Hello world")
tagger.predict(sentence)
print(sentence)
```

This, as expected, produces the same output as the model we tested right after training:

```
Sentence: "Hello world" → ["Hello"/UH, "world"/NN]
```

This shows that the words `hello` and `world` were, yet again, correctly tagged with *interjection* and *noun* tags respectively.

Resuming interrupted training sessions

While our model training example that used a severely downsampled dataset seems like a fairly straightforward process, the reality is different. Real training sessions on bigger datasets take a significantly longer amount of time to train, and the chance of unexpected interruptions therefore increases. Interrupts can happen when the machine runs out of resources because some dependency of the machine timed out, or simply because of a technical issue.

When a model training interrupt happens, all data that is stored in memory, including our trained weights, are lost.

Regardless of what really happened, the fact is, an interrupt happened and the memory-stored data is lost. Luckily, Flair has a good way around this called checkpoints. They are essentially a snapshot of the model that is stored after each epoch, regardless of whether the epoch improved the model or not. This checkpoint is stored in the `base_dir` directory with the filename of `checkpoint.pt`.

To make use of the checkpoint feature in Flair, we need to pass the `checkpoint=True` parameter value to the `trainer.train()` method when training. But the feature does come at a cost. Writing checkpoints essentially means writing model weights to the disk after each epoch, and that takes time. This means that using checkpoints is likely to slow down the model training process a bit. However, when doing serious model training for longer periods of time, this overhead is almost always worth it.

Assuming our `base_dir` is `tagger` and we have a `corpus` object-loaded memory, we can simply resume training from the last successful epoch by running the following:

```
from flair.trainers import import ModelTrainer


chkpoint_path = 'tagger/checkpoint.pt'
trainer = ModelTrainer.load_checkpoint(chkpoint_path, corpus)


trainer.train(...)
```

Here, the arguments passed to `trainer.train()` should be no different from the parameters we used for the initial training session that got interrupted, where we also used the `checkpoint=True` model trainer parameter value.

The preceding exercise and explanations should help us train a tagger for just about any language included in Flair's corpora set. But it's entirely possible that the parameters we choose will not yield good results. There are many clues in the model training output we can look out for that will tell us whether training is progressing well or whether we should restart with a different approach.

Knowing when to stop and try again

The question *"Is my model good enough?"* is a whole subfield of machine learning on its own and can't fully be explained in a single chapter. Nevertheless, there's still some practical advice we all can follow to help us determine whether it's worth restarting training with different parameters, different model types, or even new data.

The two basic techniques for measuring success during and after training are as follows:

1. Monitoring loss
2. Assessing and comparing performance metrics

Monitoring loss

The first and most important thing you should look out for during training is whether the model is learning in an expected way. This is done by monitoring training output metrics such as loss. A typical training session follows the following pattern. In the initial stage of training, our weights are randomized, and loss will be huge (remember how bad our model with random weights was?). Then, even after one or two epochs, we will see a significant improvement in loss and model performance. Finally, as we progress towards two- or three-digit epoch numbers, the improvement will get smaller and smaller. Ideally, the loss should start to plateau and converge towards a certain value. That way, assuming we didn't hit a local minimum, we can be confident that, given the parameters we chose, the model trainer did a good job. If we fail to see a convergence toward a certain value before training finishes, this may mean we finished training too soon and may need to run it for a couple more epochs.

These training debug values such as loss will be printed out during training. But these values, shown as raw numbers, are sometimes very hard to make sense of. Just by looking at the numbers, it's hard to see what the general trend is, whether the numbers are converging toward a certain value, or whether they are jumping all over the place. This is much easier to comprehend when metrics such as training loss are visualized.

Plotting training loss

Flair offers a simple tool that allows us to visually inspect metrics such as training loss and accuracy and store the output as an image.

Assuming we ran a training session from the same directory as our current Python session and used base_dir called tagger, we can plot the training loss by running the following:

```
from flair.visual.training_curves import Plotter

plotter = Plotter()
plotter.plot_training_curves('tagger/loss.tsv')
```

This will store the plot as tagger/training.png and, if you are running this code snippet inside a Jupyter notebook, display the plot in the notebook itself.

Assessing and comparing performance metrics

The final metric that will determine how good our tagger is, is the performance metric that evaluates our model on training data. In the case of PoS tagging, this will likely be the F1-score. But the value itself doesn't tell us much. For example, in the preceding exercise, we achieved an F1-score of 0.88. If we used a representative dataset, this number on its own does should give us some general understanding of how well the model will perform in the real world, but it does little in helping assess whether our tagger is something we should be happy with or whether it is better than any other tagger out there.

One of the most objective ways of assessing model performance is comparing it to the state-of-the-art results published as part of related studies or experiments. That way, if we achieved a larger F1-score, we can claim that (given the testing data that was used in evaluation) our model indeed achieves better performance. However, when comparing results across different studies or experiments, we need to be absolutely sure that the same train and test data were used in all experiments. Otherwise, we will essentially be comparing apples to oranges. For example, if one study used a dataset with very repetitive and simple language with a small vocabulary whereas the other one focused on some highly specialized and obscure uses of language with a rich vocabulary, it's safe to assume that a tagger trained on the former type of dataset will achieve better results. It's therefore vital to only compare the performance of taggers trained on same data.

Summary

The code examples, explanations, and exercises that we covered in this chapter are a quick introduction to training sequence labeling models in Flair, and it should give you the confidence to prepare, train, validate, and use Flair models that solve real-world problems.

Also, you should now be able to have the ability to tell whether model training was a success or whether it should be restarted with different parameters. We learned that there is a large number of parameters that govern sequence labeling model training as well as the training process. There's the learning rate, the number of epochs, the optimizer type, the number of hidden RNN layers, and a long list of other parameters that will likely affect the performance of our model. In the preceding examples, we usually chose the default parameters that Flair happens to use in most of its code examples, but there's no guarantee that the default parameters are the optimal parameters for our problem. In fact, it's highly unlikely. This will leave you wondering which parameters you should really use.

Luckily, Flair features a solution to this problem – a special technique that attempts model training with different combinations of parameters. It then measures performance on a special dataset called the development or validation dataset. This technique is called hyperparameter tuning. Let's learn how to do it in Flair in the upcoming chapter.

6
Hyperparameter Optimization in Flair

Grasping the concept of sequence tagging and getting a basic understanding of how it works generally isn't a huge problem. What isn't as straightforward is understanding all the parameters that govern model training and choosing the values that yield desired results. A special technique called **hyperparameter optimization** (also called **hyperparameter tuning**) helps us achieve that.

We will start with providing a general overview of what hyperparameter tuning is, why it's useful, and what different optimization methods are out there. We'll then dive into how to do tuning in Python with the **Hyperopt** library. We will conclude the chapter with a hands-on exercise where we will find the optimal hyperparameters for a **Part-of-Speech** (**POS**) tagger in Flair.

In this chapter, we will cover hyperparameter optimization in Flair as part of the following sections:

- Understanding hyperparameter tuning
- Hyperparameter tuning in Python
- Hyperparameter optimization in Flair

Technical requirements

All of the Python libraries used in this chapter are direct dependencies of Flair version 0.11 and require no special setup, assuming that Flair is already installed on your machine. Code examples covered in this chapter are found in this book's official GitHub repository in the following Jupyter notebook: `https://github.com/PacktPublishing/Natural-Language-Processing-with-Flair/tree/main/Chapter06`.

Understanding hyperparameter tuning

When first faced with a long list of model training parameters and their possible values, you might think that in order to successfully train a model, you need a special superpower that helps you pick the right parameter for the right scenario. This isn't necessarily true. While experience may help you narrow down the set of possible hyperparameters, there usually isn't a reliable way of knowing with certainty what the best hyperparameter is in advance.

Let's imagine the simplest possible scenario – a sequence tagging model trainer that receives a single parameter – say, a learning rate. This is generally a value between 0 (exclusive) and 1. To create a set of possible hyperparameter values, we simply discretize the range into a set of 10 possible hyperparameter values: `[0.1, 0.2, 0.3, 0.4, 0.5, 0.6, 0.7, 0.8, 0.9, 1.0]`. We can then perform the most trivial type of hyperparameter optimization by training 10 different models – each one trained with a different learning rate value. We evaluate the quality of each of our models and declare the highest-performing model as the one with the optimal learning rate value.

This type of hyperparameter tuning is called **grid search**, described in more detail at `https://en.wikipedia.org/wiki/Hyperparameter_optimization#Grid_search`. It is an exhaustive search algorithm, meaning it tries out the entire search space. Our preceding example illustrates the process of hyperparameter tuning if given a very dumbed down and simplified problem. Unfortunately, real-world hyperparameter selection usually isn't as forgiving for the following reasons:

- A fixed number of (in our example, 10) different learning rate values may not always be enough to find the optimal learning rate.

- The discrete uniform distribution (a distribution where the possible values are discrete and equally spaced) isn't always optimal.

- We can quickly experience the curse of dimensionality. Most real-world model trainers will govern training on more than just a single hyperparameter. This is where things get complicated. In our simplified example where we had a single hyperparameter, we had a search space of 10 possible options. Let's imagine a problem with two such hyperparameters. The number of possible parameter option combinations would be 10^2. Given three such hyperparameters, the number of possible combinations would be 10^3. This clearly suggests exponential growth that can very quickly grow out of proportion. This problem is referred to as the curse of dimensionality because with each extra parameter, we are adding another dimension that needs exploration.

Luckily, there are more sophisticated optimization methods that help us mitigate these problems.

> **Important Note**
>
> When doing hyperparameter tuning evaluation, it's best to use a dataset that is completely distinct from the data that will be used in our final model. This dataset is called a development (also known as the validation) dataset. Keeping data used for hyperparameter tuning separate from data used for training and evaluation helps us in avoiding overfitting, which would result in choosing the hyperparameters that would overfit our test dataset.

To sum up, hyperparameter tuning in NLP is simply a process that takes in a search space, an optimization method, and an objective function. It then uses the optimization (search) method to train models using a number of different hyperparameter option combinations and returns a set of hyperparameters that result in the best-performing model.

Optimization methods

Hyperparameter tuning utilizes some of the following commonly used optimization methods:

- Grid search
- Random search
- Bayesian optimization
- Evolutionary algorithms

Let's look at each of these methods in detail.

Grid search

This is a type of exhaustive search (meaning it explores the entire search space) that simply scans through all possible combinations of the predefined parameter options. It clearly suffers from the curse of dimensionality. The upside of grid search is that it can be fully parallelized.

Random search

A non-exhaustive search algorithm that simply samples random hyperparameter values and evaluates each combination. It repeats this process for a specified number of steps and returns the set of hyperparameter values that result in the best-performing model.

Bayesian optimization

A technique that uses Bayes' theorem to perform a search in order to find the optimal set of hyperparameters. It builds a probabilistic model that maps the hyperparameter value to the `objective` (loss) function.

A special variation of Bayesian optimization is the **Tree-structured Parzen Estimator (TPE)**. It is often used in the popular Python hyperparameter optimization `hyperopt` library. The library's source code and documentation can be found at `https://github.com/hyperopt/hyperopt`.

Evolutionary algorithms

Another take on hyperparameter optimization methods is the evolutionary algorithm approach. The algorithm follows these steps:

1. Generate a population of randomly chosen hyperparameter values.
2. Train and evaluate a model for each individual member of the population.
3. Replace the lowest-scoring individuals with newly formed individuals that are mutated or crossed-over variations (meaning a combination of multiple individuals) of individuals that performed well.
4. Jump to *step 2* and repeat until the desired score is obtained, no improvement is visible, or an iteration limit is reached.

We now have a basic understanding of how hyperparameter tuning works. But all these new ideas and concepts may sound a little daunting if we imagine implementing them by hand. Luckily, there are many ways to leverage the readily available libraries in Python that do just that.

Hyperparameter tuning in Python

Let's get a taste of what hyperparameter tuning looks like in practice. We will use one of the most popular Python hyperparameter optimization libraries called Hyperopt. First, let's get a general idea of how to use it in practice.

Hyperopt is a Python library that provides an easy-to-use API that requires the following three objects:

- A search space
- An objective function
- An optimization method

Let's look at each of these requirements in detail:

- **Search space** is simply the space within which the optimizer will search for different hyperparameter options. The library provides the following parameter expressions:

  - **Categorical parameters** – Parameter values that are purely categorical and can even be non-scalar (they do not need to be a number). They are provided via the `hyperopt.hp.choice` method.

  - **Integer parameters** – Integer value parameters obtained via methods such as `hyperopt.hp.quniform` or `hyperopt.hp.qlognormal`.

  - **Float parameters** – Decimal floating-point numbers obtained via methods such as `hyperopt.hp.normal` (which samples from normal distribution) and `hyperopt.hp.uniform` (which samples from uniform distribution).

- **An objective function** that determines how well a certain hyperparameter value combination performs. In deep learning, this is most often the loss function.

- **An optimization method** is an algorithm for performing a search within the search space. The options in `hyperopt` are: random search (`hyperopt.rand.suggest`), **TPEs** (`hyperopt.tpe.suggest`), and **Adaptive TPEs** (`hyperopt.atpe.suggest`).

The three preceding concepts are really all Hyperopt needs to find the optimal hyperparameters for you. Let's put this into practice as part of a hands-on exercise in Python.

Hyperparameter tuning in Python hands-on

To illustrate how hyperparameter tuning works in its simplest form, we will come up with a very simple artificial problem. We will optimize a single hyperparameter using a simple `objective` function that depends solely on that single hyperparameter, and we already know the most optimal (minimum) value for it. Let's pick this simple quadratic function:

$$f(x) = x^2 + 1$$

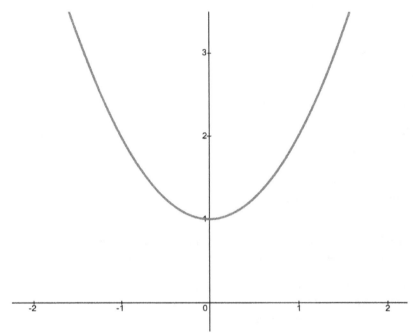

Figure 6.1 – The $x^2 + 1$ function

It's clear that the minimum of $x^2 + 1$ is 1 at $x = 0$. This now allows us to define the objective function in Python:

```
def objective(value):
    return value ** 2
```

We can then define the search space for a single hyperparameter. To keep things simple, we will choose uniform distribution and sample a range where we know the minima is, say, between -100 and 100:

```
from hyperopt import hp
```

```
# define the search space for a single hyperparameter
space = hp.uniform('param1', -100, 100)
```

The preceding code snippet will execute successfully if you have previously installed Flair. If not, make sure to run `pip3 install flair==0.11` in the console before running.

Finally, we can now choose the optimization method:

```
from hyperopt import tpe
```

```
# the Tree of Parzen Estimators (TPE) algorithm
optimization_method = tpe.suggest
```

With all three required objects defined, we can initiate the optimization process and print out the results:

```
from hyperopt import fmin
```

```
# minimize the objective function over 1000 evaluations
best_params = fmin(objective,
                   space,
                   algo=optimization_method,
                   max_evals=1000)
```

```
# print the best hyperparameter value
print(best_params)
```

The preceding code snippet prints out the following:

```
{'param1': 0.01084844145467706}
```

This indicates that, according to Hyperopt, the optimal value for our objective function is `0.01084844145467706`. This value is very close to the actual optimal value, which we already know is `0`. The obtained value is also closer to the actual optimal value (`0`) than the expected optimal values of a random search. This means that TPE did a good job of optimizing our objective function.

We now have a high-level understanding of how hyperparameter tuning works. Of course, in the real world, hyperparameters aren't used for finding the minima of simple quadratic functions. Let's find out how they're used for training sequence tagging models in Flair.

Hyperparameter optimization in Flair

As we learned in *Chapter 5, Training Sequence Labeling Models,* the success of model training often depends on a potentially large number of correctly set hyperparameters. This creates the need for finding a set of hyperparameter values that yield optimal performance. We can do this using hyperparameter optimization. Luckily, Flair offers hyperparameter tuning out of the box. Let's learn how to do it.

Hyperparameter optimization in Flair is essentially a wrapper around Hyperopt, which we briefly covered in the previous section. The extra advantage of this wrapper is that it already feeds some sequence tagging specific information into Hyperopt so that we don't have to. In the bare-bones Hyperopt coding exercise, we had to provide all three objects: search space, optimization method, and an objective function. But in Flair, we only need to define the search space and the framework will do the rest.

Hyperparameter tuning in Flair hands-on

Hyperparameter optimization in Flair is best explained through code. So let's do a simple coding exercise where we train an English PoS tagger and use Flair's hyperparameter optimization feature to find the best model's hidden size and learning rate.

We will achieve this by going through the following steps:

1. Loading the corpus
2. Defining the search space
3. Defining the `SequenceTaggerParamSelector` object
4. Setting off the optimization process

We first, just as with the sequence tagging exercise in *Chapter 5, Training Sequence Labeling Models,* need to load the English `Corpus` object into memory. It contains a tagged English corpus, split into the test, train, and dev datasets:

```
from flair.data import Corpus
from flair.datasets import import UD_ENGLISH


corpus = UD_ENGLISH()
```

```
tag_type = 'pos'

# downsample the corpus to 1% of original size
corpus.downsample(0.01)
```

In the preceding code snippet, we load the English corpus into the `corpus` variable, define our tag type as PoS tagging, and then downsample the corpus to 1% of its original size to speed up the training process for demonstration purposes.

We can then define the search space using the `hp.choice` parameter expressions:

```
from flair.embeddings import WordEmbeddings
from flair.hyperparameter.param_selection import \
   SearchSpace, Parameter
from hyperopt import hp


search_space = SearchSpace()
search_space.add(Parameter.HIDDEN_SIZE,
                 hp.choice,
                 options=[128, 256])
search_space.add(Parameter.LEARNING_RATE,
                 hp.choice,
                 options=[0.05, 0.1, 0.15, 0.2])
search_space.add(Parameter.EMBEDDINGS,
                 hp.choice,
                 options=([WordEmbeddings('en')]))
```

The preceding code snippet defines our search space, consisting of three hyperparameters. We decided to come up with some likely useful values for the learning rate and the model's hidden size. Although using parameter expressions such `hyperopt.hp.quniform` and `hyperopt.hp.uniform` would have likely yielded better results, they would also have vastly increased the search space size. Our search space is now pretty small. There are only eight possible hyperparameter combinations (two times four because there are two hidden size parameter options and four learning rate options).

> **Important Note**
>
> When doing hyperparameter optimization, some hyperparameters in
> Flair are mandatory. This includes the `Parameter.EMBEDDINGS` and
> `Parameter.LEARNING_RATE` parameters. If you do not wish to optimize
> these parameters, simply pass a single option to the search space for those
> hyperparameters.

We are now ready to define our `SequenceTaggerParamSelector` object:

```
from flair.hyperparameter.param_selection import \
    SequenceTaggerParamSelector
```

```
param_selector = SequenceTaggerParamSelector(corpus,
                                             tag_type,
                                             'paramopt')
```

This defines our hyperparameter optimization object using the corpus we loaded, the
PoS tag type that we defined, and the `paramopt` name of the folder (relative to the
directory from which we are running the code) to which the hyperparameter tuning
results will be written.

We now simply have to start the optimization process by passing in the search space and
the number of evaluations to perform. In our case, the search space is discrete and the
number option is clearly defined – 8. However, in most cases, when using non-discrete
parameter values, the number of possible hyperparameter combinations will be non-finite,
and the number of evaluations will need to be chosen empirically:

```
param_selector.optimize(search_space, max_evals=8)
```

The preceding code will train 8 different sequence taggers using 8 different
hyperparameter combinations. It will use loss as the `objective` function and evaluate it
on the development dataset. It will train each run for 50 epochs. Flair's implementation of
Hyperopt uses the TPE optimizer. The execution should take somewhere between 10 and
30 minutes on an average CPU.

When hyperparameter optimization completes, the results will be written out to
`paramopt/param_selection.txt`. The file contains the details of each evaluation
run and the following summary:

```
best parameter combination
    embeddings: 0
```

```
hidden_size: 1
learning_rate: 3
```

The preceding summary indicates that the best hidden size is the second option (because the summary says 1 and we use zero indexing, meaning 1 refers to the second option): 256. The best learning rate is the fourth option – 0.2. For embeddings, of course, because we only had a single option to choose from, the first and the only option was chosen as the best.

Does this mean that we should always use 256 as the hidden size and 0.2 as the optimal learning rate for English? In short, no. Optimal hyperparameter values will depend on the data used, the problem attempting to be solved, and the techniques used to tackle it. Remember the preceding code snippet where we loaded the corpus and reduced it from 100% to 1% of its original size? By doing this, we greatly increased the speed of model training, which allowed us to perform hyperparameter optimization much quicker, but by doing so, we also focused our model on a very small and potentially biased subset of data. To perform proper hyperparameter tuning on an English dataset, we'd need to do it at 100% corpus size and consider more hyperparameter options. This would vastly increase our search space and execution time, but also improve the performance of our model. These findings lead us to the following takeaway.

Summary

Hyperparameter tuning is an utterly useful technique for finding optional hyperparameters, not just in sequence tagging, but across many areas of machine learning. However, it does so at a cost; it is incredibly expensive. In essence, it performs the training and evaluation process tens, hundreds, or potentially thousands of times. But in the real world, exploring each and every potentially useful hyperparameter value can take months to execute, and the search space is usually narrowed down to a finite set of potentially useful options. Hyperparameter tuning is, therefore, a process that does require a certain level of expert knowledge and experience that will help you achieve the correct trade-off between exploration (trying out different not-yet-explored options) and exploitation (narrowing down the parameter values to the ones that generally seem to work well).

As part of this chapter, we learned why hyperparameter tuning is useful, when to use it, and how. We learned how to use a hyperparameter tuning library in its purest form in Python as well as how to use it through Flair to tune sequence-labeling models.

In the next chapter, we will be covering yet another NLP problem where hyperparameter tuning proves to be useful – training embeddings.

7

Train Your Own Embeddings

The reason Flair sequence taggers yield such outstanding performance can mainly be attributed to its secret sauce – Flair embeddings. Their contextual design, the fact that they are character-based, and the way they can be used in the backward-forward configuration make them a perfect fit for use in sequence labeling tasks. But, up until this point in this book, we haven't focused much on how these embeddings are actually trained.

In the previous chapter, where we covered model training and learned about word embeddings, we simply used the pre-trained embeddings that were available as part of the Flair Python package. But there are many **Natural Language Processing** (NLP) problems we may stumble upon where the pre-trained embeddings will not be sufficient.

When working with flair, you may find yourself dealing with a language that isn't covered by Flair's pre-trained embeddings yet, or you may simply need embeddings that are trained on some specific corpus. When this happens, you can make use of Flair's built-in classes that allow you to train your own Flair embeddings, which can later be used to train sequence taggers or solve other downstream NLP tasks.

We'll begin with explaining why and in what circumstances training custom embeddings may be useful. We'll cover how Flair embeddings work and how they achieve such great performance on downstream sequence labeling tasks. We will then focus on methods for evaluating word embeddings. Finally, we'll train our own embeddings as part of a hands-on exercise on the world's smallest language.

In this chapter, we will cover training Flair word embeddings as part of the following sections:

- Understanding the how and why behind custom Flair embeddings
- Evaluating word embeddings
- Training Flair embeddings on the world's smallest language

Technical requirements

The exercises in this chapter require Flair version 0.11, as well as the `requests` Python package, to be installed on your development machine.

The code examples covered in this chapter are found in this book's official GitHub repository in the following Jupyter notebook: `https://github.com/PacktPublishing/Natural-Language-Processing-with-Flair/tree/main/Chapter07`.

Understanding the how and why behind custom Flair embeddings

Word embeddings play an important, if not essential, role in sequence tagging models' performance. The details of what embeddings are, is covered in *Chapter 3*, *Embeddings in Flair*. But, for the purposes of this chapter, it's important to understand that embeddings are essentially word representations most often found in the form of real-valued vectors. These vectors can then be used as input on a number of downstream tasks, such as **part-of-speech** (**PoS**) tagging and **named entity recognition** (**NER**). Let's first quickly cover how Flair generates embeddings and how they are trained.

Why training embeddings rarely ever means training embeddings

The term **training embeddings** is very often a confusing term that dates back to the older methods, where the result of training embeddings was a set of word embeddings. This term makes less sense with Flair's way of training embeddings, and even less so given that the embeddings are contextual and character-based. For the more traditional types of embeddings, such as **GloVe**, once the embeddings are trained, we can simply memorize an embedding for each word in the dictionary. At this point, someone who doesn't fully understand the design of Flair embeddings might wonder the following:

Given that Flair's embeddings are contextual, doesn't this mean that the process of training embeddings results in precomputing one embedding for each word in the dictionary for each possible context we may find the word in?

The answer is no. With word-based, non-contextual word embeddings, the number of possible embeddings is indeed bound by the number of words in the dictionary. But, because Flair embeddings can model virtually any sequence of characters (not necessarily a known word) in any given context, the number of possible embeddings is unimaginably large. This is because of the following:

- Flair embeddings are contextual, meaning the same word will result in a different embedding when put in a different context.
- Flair embeddings are trained without an explicit notion of words. Words are modeled as sequences of characters.

When we talk about training Flair embeddings, we are not really talking about pre-computing the embeddings themselves. Instead, we are talking about training a language model. Let's learn more about that.

The idea behind Flair embeddings

Training Flair embeddings is really a process of training language models. Flair embedding models are essentially language models that predict the next character given some context (surrounding text). Flair uses language models based on **Long Short-Term Memory (LSTM)** recurrent neural network nodes. These nodes are capable of processing sequential data.

> **What Is a Language Model?**
>
> In simple terms, a language model is a tool that models a probability distribution over strings of text. It assigns a probability to each possible character, word, or sequence of words often given some context (preceding text). A popular type of language model, up until the emergence of deep learning, was the statistical n-gram model; but, the more recent neural-based approaches consistently outperform the n-gram models. There are still some advantages to the classic n-gram models. The n-gram models, unlike most neural models, don't require much data and can be very light in terms of size.

Character-based language models receive input text as a sequence of characters and are trained to predict the next character. The model is then trained so that it learns to predict the probability distribution of the possible next characters given past characters.

So, what does predicting the next character have to do with word embeddings?

A trained Flair language model in essence *is* the model for producing word embeddings. First, we pass the entire context (text that surrounds our word), including the word we want to compute the embedding for, to the model as input. Then, instead of focusing on the predicted character, we focus on a special value of the LSTM cell called the **hidden state**. We extract the hidden state after the last character in the word, and this is our word embedding.

It is clear that context affects the semantics of certain words. But, you may wonder whether it's the text that *precedes* a word that influences the meaning of the word, or whether it's the text that comes *after* the word that influences its meaning. The answer is *both*.

The forward-backward embedding stack

One brilliant design concept in Flair embeddings that helps them achieve their great performance is the way they use a backward-forward embedding stack.

The **forward model** takes into account the entire context before the word and the word itself. It captures the semantic meaning of the word influenced by the text that precedes the word. But, this model doesn't really take into account all the words that come after the word we're trying to compute the embedding for. To accomplish this, we would need to build a model that works in a reverse direction. We would need to build a **backward model**.

A backward model processes words in the opposite direction. For English, this would mean from right to left. It takes into account all the context after the word plus the word itself. This model captures the semantic meaning of the word influenced by the text that follows the word.

Once both models are trained, we can then compute a forward embedding and a backward embedding. We then usually join them together to form a new embedding that we call the **embedding stack**.

Corpora for Flair embeddings

Training any type of machine learning model requires data. Luckily, training embeddings is an unsupervised learning task. The data used for training is unlabeled. What this means is that potentially any text that contains meaningfully constructed words can be used as part of our corpora. What is important is that we use a large enough corpus.

There is also the question of copyright. There is an ongoing legal question about whether the copyright-related rights of the corpus also apply to the language model itself, and under what conditions. For example, if attribution is required to distribute an original corpus, do the same attribution requirements also apply to a language model that was trained on that same corpus? The answer to this question depends on many factors, but there is currently no clear way of answering in a way that would apply to all types of scenarios. But, some general rules still do apply. If you are learning about NLP at home on your personal machine, it's unlikely you will get yourself into trouble if you train your model on any publicly available text. These practical exemptions, however, do not apply when working in a commercial environment. The enforcement of the copyright laws gets progressively stricter as you move toward the larger tech companies.

Dictionaries for training Flair embeddings

One of the best things about Flair embedding training is the effort, or lack thereof, needed to prepare a dictionary for the language model. With most traditional word-based language models, you would need to come up with a complete list of words for a given language.

Luckily, Flair embeddings are character-based. This means that the input to the language model isn't words, but rather sequences of characters. We need not worry about coming up with a complete list of words. Instead, we only need a list of all possible characters used in our language. This means that if we are dealing with a language that uses Latin script, our dictionary will contain only a handful of items. Flair also provides classes with predefined commonly used dictionaries, such as Latin.

Flair's pre-trained embeddings yield state-of-the-art results on many downstream NLP tasks in many languages. But, when training new embeddings models, the performance will depend on many factors, and not every embedding model we train will perform well. We need to find out ways of figuring out whether our newly trained embeddings are up to par.

Evaluating word embeddings

In the previous section, we covered the design of Flair embeddings that use language models. The process of training these language models isn't much different from any other type of deep learning training. But well-performing language models don't necessarily mean good embeddings that yield excellent results on downstream tasks.

Instead, there we typically need to rely on the following two approaches of evaluating word embeddings:

- **Intrinsic evaluation** aims to test the quality of embedding word representations independent of any natural language processing tasks. This is done by measuring semantic relationships between words. The simplest type of intrinsic embedding evaluation is word similarity. It simply uses a similarity metric, such as cosine distance, to measure the similarity between word embeddings and compares it to the human-perceived semantic similarity. For example, the words *begin* and *start* are semantically very similar. If our embeddings are well trained, we would then expect those two words to be close together in cosine distance.

  Other types of intrinsic evaluation methods exploit the analogy relationships between words, similar to how we learned about word embeddings as part of the *king is to man as queen is to woman* analogy exercise in *Chapter 3, Embeddings in Flair*.

- **Extrinsic evaluation** aims to test the quality of word embeddings by using them in a downstream NLP task such as PoS tagging, NER, or chunking. We feed the embeddings as input to an NLP model. We then evaluate the quality of the NLP model and assume its improvement/regression in performance can be attributed to the performance of our embeddings. Flair has all the tools for training sequence labeling models, as well as offering labeled datasets for commonly spoken languages. If you find yourself training Flair embeddings, chances are you are planning to use them as input to a downstream sequence tagging NLP task. This often makes extrinsic evaluation a straightforward choice.

Let's now jump on to the next part of this chapter, where we train our own embeddings, and later, perform evaluation using one of the techniques described here.

Training Flair embeddings on the world's smallest language

The Flair package provides a simple, straightforward API for training Flair embeddings. The process involves all the necessary steps required to train any language model. The steps include the following:

- Preparing the dictionary
- Preparing the corpus
- Defining the language model
- Training the language model

Let's cover the process of training Flair embeddings through a practical hands-on exercise.

Training embedding for most languages is not a quick process. A decent GPU-equipped machine would require over a week of training time to produce results comparable to the state-of-the-art published Flair results in English or similar languages. Most pre-trained Flair embeddings, such as the *en* English embeddings model, produce embeddings of length 2,048.

Part of the reason why we need 2,048 dimensional vectors is that languages such as English have a huge number of words in the dictionary, and their semantics change based on the context in which they are used. But what if we, as part of this exercise, trained embeddings on a very small language? What if we trained embeddings on a language so small, that we could do it even on a machine without a dedicated **graphics processing unit (GPU)**?

As part of this exercise, we will train Flair contextual word embeddings for the world's smallest language – **Toki Pona**.

Toki Pona is a constructed (made up) language known for its tiny vocabulary. The original version of the language has only around 120 words. The language was created around the concept of minimalism with the aim of simplifying thoughts and communication. It has only has nine consonants and five vowels, and the pronunciation of Toki Pona words is compatible with most world's languages.

Let's now train a tiny Flair embeddings model for Toki Pona.

Preparing the dictionary

When training classic word embeddings, the process of defining a complete dictionary for a language can often be rigorous and exhausting. This isn't the case with Flair embeddings.

It's because Flair embeddings model characters, not words. This means that in order to define a dictionary, we don't need a complete list of all possible words. We instead only need a list of all possible characters.

In the Toki Pona language, only the following 14 characters are used: `ptksmnljwaeiou`. These characters are a subset of the Latin character set. Flair's pre-compiled Latin dictionary can simply be loaded by running the following:

```
from flair.data import Dictionary

dictionary = Dictionary.load('chars')
```

But, it's important to know that this character set contains many characters that are never used in Toki Pona. This would mean that the language model would need longer to train in order to learn that some of the characters defined in the dictionary are meaningless.

Instead, if we want to keep our dictionary minimal and only include the letters that are used in our language, we can construct our own `Dictionary` class manually by using the `Dictionary.add_item(c)` method:

```
from flair.data import Dictionary

dictionary = Dictionary()
toki_pona_symbols = 'ptksmnljwaeiou'
toki_pona_symbols += toki_pona_symbols.upper()

for c in toki_pona_symbols + '?. ':
    dictionary.add_item(c)
```

In the previous code snippet, we defined a dictionary that contains all possible characters used in Toki Pona in both upper and lowercase, along with the space and two punctuation symbols. The `Dictionary` class also automatically adds a special <unk> symbol, used to mark unknown characters.

Preparing the corpus

Without data, it's impossible for our embeddings model to learn anything. The general rule with training embedding models is that the more data, the better. In practice, this means that a good embedding model will require well over a million sentences. But, for the purposes of this exercise, and because we are dealing with a language with roughly 120 possible words, we can pick a much smaller corpus. We will pick a publicly available collection of 1,000 sentences in Toki Pona available via this address: `https://raw.githubusercontent.com/tadejmagajna/TokiponaCorpus/master/bot4/corpus.txt`.

We will then split the corpus into a train, test, and validation dataset and pass the corpus directory to a `TextCorpus` class constructor.

Let's first download the corpus and split it into test, train, and validation sets:

```
import requests

url = "https://raw.githubusercontent.com/tadejmagajna/
TokiponaCorpus/master/bot4/corpus.txt"
response = requests.get(url)

sentences = response.text.splitlines()
one_tenth_corp_len = int(len(sentences)/10)

test, valid, train = (
    sentences[:one_tenth_corp_len],
    sentences[one_tenth_corp_len:one_tenth_corp_len*2],
    sentences[one_tenth_corp_len*2:])
```

The previous code snippet first downloads the entire corpus from GitHub. It then splits the sentences by new lines and loads the first 10% of the corpus into `test`, the other 10% of the corpus into `valid`, and the remaining 80% of the corpus into `train`.

It is also worth mentioning that dataset splitting is in practice almost always done using helper NLP framework functions. Those helper functions can assure the dataset is properly shuffled so the data is well distributed between sets. But, in this code example, we can assume the data is distributed well already, meaning we can simply slice the list of sentences into three parts.

Important Note

To run the preceding code snippet, you will need a local Python 3 installation with `Flair` version 0.11 along with the `requests` package. The `requests` package can be installed by running `pip3 install requests`.

We now have the corpus stored in memory (which wouldn't be possible if we dealt with a corpus with millions of sentences). Before we initialize the `TextCorpus` class, we need to store the corpus sets in a local directory that contains the corpus organized into the following folders and files:

- A `valid.txt` file that contains the validation set

- A `test.txt` file that contains the test set

- A `train/` folder that contains the train set split into numbered files that follow the following naming pattern: `train_split_1`, `train_split_2`, `train_split_3`, and so on

We organize the corpus into the structure defined previously by running the following code:

```
from tempfile import TemporaryDirectory
from os.path import join
from os import mkdir


dataset_dir_obj = TemporaryDirectory()
dataset_dir = dataset_dir_obj.name
train_dir = join(dataset_dir, 'train')
mkdir(train_dir)


with open(join(dataset_dir, "test.txt"), "w") as file:
    file.write(' '.join(test))
```

```
with open(join(dataset_dir, "valid.txt"), "w") as file:
    file.write(' '.join(valid))

with open(join(train_dir, "train_split_1"), "w") as file:
    file.write(' '.join(train))
```

In the first part of the preceding code snippet, we use the `tempfile` library to create a temporary directory in which we will store our corpus. Using temporary directories in Python is a good idea because the library provides an easy interface to clean up after we're done.

In the second part of the preceding code snippet, we then write `test`, `valid`, and `train` corpus sets into `train.txt`, `test.txt`, and `train_split_1` files respectfully. The entire corpus is now stored in the `dataset_dir` directory.

Now that we have the corpus organized properly, we can initialize the `TextCorpus` class:

```
from flair.trainers.language_model_trainer import TextCorpus

corpus = TextCorpus(dataset_dir,
                    dictionary,
                    forward=True,
                    character_level=True)
```

The preceding code snippet initializes the `TextCorpus` class where we provide the dataset directory with the dictionary, and we explicitly tell the constructor that we are preparing a corpus for a forward model (one that takes into account the word and the context that precedes it).

Training the language model

We can now define and train our language model:

```
from flair.models import LanguageModel
from flair.trainers.language_model_trainer import (
    LanguageModelTrainer)

language_model = LanguageModel(dictionary,
                              is_forward_lm=True,
                              hidden_size=64,
```

```
                              nlayers=1)

trainer = LanguageModelTrainer(language_model, corpus)
trainer.train('forward_model_directory',
              sequence_length=25,
              mini_batch_size=10,
              max_epochs=100)
```

In the first part of the preceding code snippet, we initialize the `LanguageModel` class by telling it that we are building a forward model with a single layer and a hidden size of 64. `hidden_size` is an important parameter that defines the size of the LSTM nodes' hidden size, which in effect defines the dimensionality of our embedding vectors. `hidden_size` would generally need to be set much higher, to a number such as 2,048. But, because we are training a model on a language with roughly 120 words and we are doing this only as part of an exercise, we can get away with low-dimensional embeddings.

In the second part of the previous snippet, we start the model training process by initializing the `LanguageModelTrainer` class and telling it to use a sequence length of 25 and run it for 100 epochs. The parameters chosen here are purposely much lower than the parameters that would have been used if we were training a large embedding model for production. In practice, it's a good idea to use longer sequence lengths than this and do more epochs.

We then start training by calling `LanguageModelTrainer.train()` where we specify the output directory of the trained model, which in our case is `forward_model_directory/`. This directory will be created relative to the directory we are running this code from.

If we did everything right, the model training should commence, and Flair should start spitting out some debug text and metric values to standard output:

```
ms/batch   5.88 | loss   0.77 | ppl      2.15
Epoch      97: reducing learning rate of group 0 to 1.2207e-03.
best loss so far   0.98
('\n ni li sike. mi pini e ma mije? kala ni li sin e pan lawa
en kena t
-------------------------------------------------------------------
end of split 1/1 | epoch  97 | time:   1.75s | valid loss 0.99
-------------------------------------------------------------------
Epoch time: 1.92
read text file with 1 lines
```

The preceding output provides valuable information. It tells us which epoch we are processing at the moment (epoch 97 in this example), what the train loss is (0.77 in this example), what the validation loss is (0.99 in this example), and it prints out some text generated by our language model: ni li sike. mi pini e ma mije? kala ni li sin e pan lawa en kena t.

In the first few epochs, you will notice that our language model is producing gibberish. Most words are out-of-vocabulary words and the output is completely random. But, as training progresses, you will notice the loss decreasing and the generated text starting to look like Toki Pona.

Training these 100 epochs on a CPU shouldn't take more than 10 minutes on an average machine. When training finishes, Flair will print out the final validation loss and perplexity:

```
TEST: valid loss  0.92 | valid ppl 2.52
```

And that is it!

Our Flair embedding model is now trained. We can start generating some embeddings. But, before we do that, we can play around a little with the language model and generate some text using the generate_text() method:

```
t = language_model.generate_text(number_of_characters=40)[0]
print(t)
```

This will print out 40 characters generated by the language model (results will be different for each training run because of randomly initialized weights):

```
sike ni li pona mute. soweli ni li pana
```

While these Toki Pona sentences make little sense to a human, at least we can be pleased that the words generated are actual Toki Pona words and the word structure seems to follow *some* sort of a logical pattern. Do also note that we only generated 40 characters, which means that the last word can potentially be cut off. If you train your model yourself, your randomly initialized model weights will be different and the text your model produces will almost certainly be different from the example shown.

Using custom embeddings on downstream tasks

The primary purpose of training the Flair embedding model isn't playing around with generated text and printing out embedding vectors. Chances are, we are training embeddings so that we can use them on a downstream NLP task.

Luckily, the syntax for loading custom-trained Flair embeddings is identical to loading any other pre-trained Flair embedding model. We simply need to pass the model path to the constructor.

If the output path passed to the `LanguageModelTrainer` class was `forward_model_directory/`, we will find our trained model at `forward_model_directory/best-lm.pt`.

We can then instantiate a `FlairEmbeddings` object by running the following:

```
from flair.embeddings import FlairEmbeddings


fw = FlairEmbeddings('forward_model_directory/best-lm.pt')
```

The preceding code snippet loads our custom embedding model into the `fw` variable.

Note that in this exercise, we only trained the forward embedding model, which takes into account the context before the word. For added performance, you may now want to repeat the entire embedding training process with the `forward` and `is_forward_lm` parameters set to `False`, which will train the backward model instead. You can then stack the embedding models together to form a `StackedEmbeddings` object, a part of the process covered in depth in *Chapter 5, Training Sequence Labeling Models*.

You can now use this model on any Flair downstream NLP task that uses word embeddings. Doing this will also allow you to perform extrinsic evaluations of the embedding models by evaluating the performance of taggers that use these custom embeddings.

But training taggers takes a long time. So, before you do just that, you may simply want to do a sanity check to verify our embeddings carry at least some useful information.

Performing intrinsic evaluation on custom Flair embeddings

Before we go through the lengthy process of training sequence tagging models to verify whether our embeddings make any sense, we can instead perform a very simple intrinsic evaluation as a sanity check.

A simple intrinsic check we can do is to compute embeddings for a few synonyms, and then use a similarity metric to determine whether the resulting embeddings are similar. Similarly, we can compute embeddings for words that are semantically dissimilar and check whether their embeddings are dissimilar as well. To make our lives easier, we will use an **out-of-vocabulary** (**OOV**) word for the dissimilar words. Our test will, therefore, verify whether embeddings of two similar words are closer together (in terms of cosine distance) than embeddings of one dictionary word and one OOV word.

For similar words, we will choose the words `lukin` and `oko`. While not actual synonyms, their meaning in Toki Pona relates to vision and eyes. We will generate embeddings for these words and one other OOV word, and then measure the similarity between those embeddings. If our embeddings are any good, `lukin` and `oko` should score high on the similarity metric, whereas the OOV word is not expected to be similar to the previous two.

Let's compute embeddings for all three words using our custom embedding model:

```
from flair.embeddings import FlairEmbeddings
from flair.data import Sentence

synonym_1 = Sentence('lukin')
synonym_2 = Sentence('oko')
rand_word = Sentence('oovword')

fw = FlairEmbeddings('forward_model_directory/best-lm.pt')
fw.embed(synonym_1)
fw.embed(synonym_2)
fw.embed(rand_word)

embedding_syn_1 = synonym_1[0].embedding.tolist()
embedding_syn_2 = synonym_2[0].embedding.tolist()
embedding_rnd_wrd = rand_word[0].embedding.tolist()
```

In the first part of the preceding code snippet, we simply initialize the `Sentence` classes for the words `lukin`, `oko`, and `oovword` (the random OOV word).

We can then use the cosine similarity metric to measure the similarity between those three embeddings:

```
from sklearn.metrics.pairwise import cosine_similarity as sim

s_synonym = sim([embedding_syn_1], [embedding_syn_2])[0][0]
s_rand_1 = sim([embedding_syn_1], [embedding_rnd_wrd])[0][0]
s_rand_2 = sim([embedding_syn_2], [embedding_rnd_wrd])[0][0]
```

The preceding snippet computes the similarity between the three word pairs and stores the similarity between `lukin` and `oko` (the synonyms) into `s_synonym`. It also computes the similarity between `lukin` and `oovword`, and `oko` and `oovword`.

Let's see if our assumptions were correct:

```
print(s_synonym , s_rand_1, s_rand_2)
```

The preceding code snippet, if you trained your embeddings well, will print out something similar to the following:

```
0.6111595281781783 0.2691232575537895 0.18826418665007352
```

Hooray!

Turns out our hypothesis was supported. The similar words scored highly in the similarity score (the first number printed), whereas the unrelated words (the second and third number printed) scored lower. Note that the only thing we may conclude from these results is that there is *some* level of Toki Pona information stored in our embeddings, but we shouldn't jump to any conclusions on how useful these embeddings would be on downstream NLP tasks. To train Flair embeddings properly for real-world use, we would need to train them on a much larger dataset, using a higher embeddings dimensionality, for more epochs, and using a longer sequence length.

Summary

Training custom embeddings is one of the more complicated Flair features. It requires all the knowledge and understanding that helps us choose the right parameters, and prepare data correctly to make use of the huge compute power required to train embeddings on larger languages. It is also one of the key concepts to understand and perform correctly because almost everything else that Flair does depends on embeddings in one way or another.

In this chapter, we covered the motivation behind training custom Flair word embeddings and did an overview of the embeddings design. We covered the syntax required to train these embeddings by training forward word embeddings for the world's smallest language – Toki Pona.

So far, we have presented embeddings as something that can be used as an input to a downstream NLP sequence labeling task. But, embeddings can also be used in other NLP applications, such as text classification. Let's learn about that in the next chapter.

8
Text Classification in Flair

After covering complex topics such as hyperparameter tuning, embedding and sequence labeling model training, we're now ready to move to a slightly more straightforward and easy to grasp part of **Natural Language Processing** (**NLP**). In this chapter, we'll be covering **text** (also known as **document**) **classification**. While Flair's strength traditionally lies in **sequence tagging**, the library offers solutions that leverage both Flair embeddings as well as other third-party solutions that allow it to yield decent results with text classification. Some of its strengths lie in its simplicity in training while others lie in the zero and few-shot classifiers – classifiers that require little to no training data.

We'll be starting off with some background on what text classification is, what it can be used for, how it is trained, and how success is evaluated. We'll later move on to what Flair can offer for text classification and cover the different types of text classifiers available. We will train our very own classifier as part of a hands-on exercise. Finally, we will explore a novel approach to text classification that requires little to no training data.

We'll be learning about text classification and how to train and use text classifiers in Flair as part of the following:

- Understanding text classification
- Text classification in Flair

- Training a text classifier in Flair
- Working with text classifiers that require little to no training examples

Let's dive in!

Technical requirements

All of the Python libraries used in this chapter are direct dependencies of Flair version 0.11 and require no special setup, assuming Flair is already installed on your machine. Code examples covered in this chapter are found in this book's official GitHub repository in the following Jupyter notebook: `https://github.com/PacktPublishing/Natural-Language-Processing-with-Flair/tree/main/Chapter08`.

You will be able to run most of the code and train most models in this chapter using a CPU. However, the final exercise does require a GPU-equipped machine.

Understanding text classification

Text or document classification is simply a process of assigning one or more labels (often called **classes**) to a piece of text (often called a document). A text classifier is a machine learning model that receives some text as input and computes a probability distribution over a set of classes.

Text classification has many real-world uses and is actively used to solve the following problems:

- **Topic classification** – the process of assigning a topic to a document
- **Spam detection** – the process of detecting unwanted emails
- **Sentiment analysis** – classifying text sentiment into positive and negative
- **Hate speech detection** – identifying hate speech in text
- **Language identification** – figuring out what language a document is written in

Here's an example of the real world use of text classification:

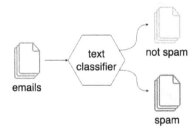

Figure 8.1 – Spam detection

In relation to how documents are classified, we know three different types of text classification:

- **Binary classification** – there is only one possible class in binary classification. Each document either belongs to that class or doesn't belong to that class.

- **Multiclass classification** – only one of the many possible classes is assigned to a document.

- **Multilabel multiclass classification** – one document can belong to multiple classes simultaneously.

It should now be clear that text classification, while a fairly simple concept to grasp, is still an important part of NLP. But before we dive straight into how to classify text in Flair, let's have a quick look at how text classifiers are trained in general.

Training text classifiers

Document or text classification is a form of supervised machine learning. This means that a machine learning model is trained on labeled data that acts as the ground truth. For example, for a spam detection classifier, each document would be labeled as *spam* or *not-spam*.

There are no rules in text classification that define what and how many classes can be defined to solve a particular problem. The set of possible classes is usually artificially constructed so that it solves a certain business problem in the most optimal way. Text classification is therefore different from other forms of supervised machine learning, in the sense that there is usually no objective ground truth other than the labeled data itself. For example, for part-of-speech tagging, the ground truth, the solutions to how the words are supposed to be tagged, is defined by grammar, and there is a clear answer to what part of speech each word in the sentence should be classified into. But with text classification, the classes are often abstract, high-level concepts with no clear definitions or boundaries. Class assignments are, therefore, often subjective. For example, in sentiment analysis, we need to classify sentences into *positive* and *negative*. However, different people will have different opinions about whether certain sentences are positive or negative. Therefore, with classification problems, often there is no ground truth other than the labeled data itself. This is one of the reasons why text classifiers are susceptible to labeling bias. It means that if the class labels in our training text contain bias, our predictions, regardless of how well we train the model, will be biased.

Detecting and mitigating bias is a complex problem that extends NLP and, even more so, extends the scope of this book. Here, we plan to cover how the models are trained and, more importantly, once trained, how and what they are evaluated against.

The process of training a text classifier is not a whole lot different from training sequence taggers. We start off with a certain deep learning model with randomly initialized weights, and then we train the model in a way that fits the training data. Of course, we don't want our model to fit the training data too well in an effort to avoid overfitting. This is why it's important to assess the model on the test dataset. To evaluate the model and determine whether it works as intended, we use one of the following metrics.

Metrics for single and multi-label text classification

The simplest and easiest way to understand a metric that can be applied to many classification problems including text classification is **accuracy**. Accuracy is simply the proportion of correctly classified data. It can be computed using the following formula:

$$\frac{correctlyClassified}{(incorrectlyClassified + correctlyClassified)}$$

Here, `correctlyClassified` is the number of examples that were assigned the correct label, and `incorrectlyClassified` is the number of examples that were assigned an incorrect label. While this metric results in a number between 0 and 1, it can be multiplied by 100 and expressed by a percentage (%) value.

For text classification problems with balanced class distribution (where there is the same number of examples that belong to each class in our training data) and single-class classification, this metric is often as good as it gets. But when we start dealing with multi-label (assigning more than one class to each training example) classification or imbalanced datasets, that's when we may look further toward other metrics, such as the **F1 score**.

The F1 score was already well explained in *Chapter 4, Sequence Tagging*, and the same F1 score concepts that apply to sequence labeling there apply to text classification here. The score is a harmonic mean of two other metrics called **precision** and **recall**. It ranges from 0 (worst possible score) to 1 (best possible score).

To fully understand the F1 score in terms of classification, we need to assess each class label individually. For each predicted label, we check whether it is the label in question (meaning positive) or not (meaning negative). We then compute the following metrics.

In classification, `precision` is the proportion of positively labeled examples that are truly positive:

$$precision = \frac{tp}{tp + fp}$$

And `recall` is the proportion of positive examples that were actually correctly classified:

$$recall = \frac{tp}{tp + fn}$$

Where:

- **True Positives (TP)**: The number of examples predicted positive that are actually positive

- **False Positives (FP)**: The number of examples predicted positive that are actually negative

- **False Negatives (FN)**: The number of examples predicted negative that are actually positive

The F1 score is then defined as the harmonic mean of precision and recall:

$$F1 = 2 \cdot \frac{precision \cdot recall}{precision + recall}$$

The process of working out the F1 score is the same as the one used for sequence labeling and involves the following steps:

1. Counting true positives, false positives, and false negatives
2. Getting precision and recall using the preceding formulae
3. Computing the F1 score as the harmonic mean of precision and recall

If we compute the F1 score for each class label separately and then average all the F1 scores, we obtain the **macro F1 score**. This metric takes into account imbalanced class distributions in our dataset.

If we compute the F1 score for all the classes counting the total true positives, false negatives, and false positives together, we obtain the **micro F1 score**. This metric will have the same value as accuracy in single-label classification.

Let's see how this theory holds up in practice with one of the simplest yet most powerful NLP libraries out there.

Text classification in Flair

Flair offers a simple interface for using pre-trained models as well as training custom text classifiers. While Flair's secret sauce – the forward and backward Flair embeddings – aren't the best tool for text classification tasks, Flair can leverage other third-party methods to yield excellent performance in text classification.

Let's first learn how to train and use Flair's pre-trained models.

Using pre-trained Flair text classification models

The set of pre-trained text classification models available in Flair is fairly small, and chances are, the model you're looking for isn't there. Even so, the following syntax will be valuable, as the API for loading and using the Flair pre-trained models is the same as the API for loading and using custom-trained models. We will train those in the next section.

Flair currently offers the following stable pre-trained models:

- `sentiment` – a sentiment analysis model for English to classify text as positive and negative

- `sentiment-fast` – a slightly smaller version of the sentiment text classification model with faster inference

- `de-offensive-language` – a text classification model for labeling German text as either offensive or non-offensive – for example, the model will label sentences that contain profanities as offensive

Knowing which text classification model names are currently available in Flair, we can now load and use a model using the following syntax:

```
from flair.models import TextClassifier
from flair.data import Sentence

classifier = TextClassifier.load("sentiment")
sentence = Sentence("Flair is pretty cool!")
classifier.predict(sentence)

print(sentence.labels)
```

The preceding code snippet will first load a classifier using the `TextClassifier.load()` function by passing in the `sentiment` model name. The first time we load this model, Flair will download it from its web servers onto our local machine. These models will then be cached on the disk for future use.

We then initialize a `Sentence` class with a sentence that clearly delivers a positive message. Next, we will use the `TextClassifier.predict()` method to generate predictions and store them as part of our `Sentence` object.

Finally, we predict the class label probability by simply printing out the object. It produces the following output:

```
['Sentence: "Flair is pretty cool !"'/'POSITIVE' (0.9991)]
```

The preceding message shows that our `Flair is pretty cool!` sentence was assigned a single `POSITIVE` class label. The message also indicates that our model is 99.9% confident in its decision.

We can try the exact opposite with a profoundly negative sentence:

```
sentence = Sentence("I don't get enough sleep.")
classifier.predict(sentence)


print(sentence.labels)
```

This prints out the following:

```
['Sentence: "I don't get enough sleep ."'/'NEGATIVE' (0.9911)]
```

This indicates that our model is 99.1% sure that the preceding message has a negative connotation.

And that's it! A few lines of code and we can already classify text. The downside of Flair's pre-trained text classification models is that there's sadly only a handful to choose from. Luckily, Flair provides an easy-to-use interface to train new custom models, and we can train as many as we want! Let's learn how to do that by first learning about how to leverage the underlying technology required to perform text classification in Flair – document embeddings.

Document embeddings in Flair

The same way word embeddings are a key part of producing sequence labeling models in Flair, many text classification models leverage a similar concept – document embeddings.

Document embeddings are essentially vector representations of documents. The term *document* simply refers to a sequence of words. The length of these documents can range from a single word to an entire book or corpus.

Flair offers a set of different types of document embeddings as discussed in the following subsections.

DocumentPoolEmbeddings

This is the simplest type of document embeddings. It is a piece of code that computes word a word embedding for each word in the document and returns an average. For example, if we had a two-word document and the first word's embedding was [1, 2] and the second word's embedding was [3,0], the DocumentPoolEmbeddings embedding would be [2, 1]. Other than computing the average embedding (referred to as mean pooling), this type of document embedding can also use other means of pooling, such as min pooling (meaning it takes the smallest value for each embedding dimension) and max pooling (meaning it takes the maximum value of each embedding dimension). To use DocumentPoolEmbeddings, you only really need the trained word embeddings. The document embeddings themselves require no training, which is a huge plus!

```
from flair.data import Sentence
from flair.embeddings import (DocumentPoolEmbeddings,
                              WordEmbeddings)

glove = WordEmbeddings('glove')
document_embeddings = DocumentPoolEmbeddings([glove])
sentence = Sentence('two words')
glove.embed(sentence)

print((sentence[0].embedding + sentence[1].embedding) / 2)
```

In the preceding code snippet, we implemented document embeddings with mean pooling ourselves. We simply loaded a word embedding object and computed word embeddings for two words (the word two and the word words). We then summed the two word embeddings up and divided the resulting embedding by two – thus averaging the two embeddings or performing mean pooling. We can do the exact same by using the DocumentPoolEmbeddings class:

```
document_embeddings = DocumentPoolEmbeddings([glove])
document_embeddings.embed(sentence)

print(sentence.embedding)
```

The preceding code snippet uses the DocumentPoolEmbeddings class to compute the embedding for our document. The output of printing out the embedding shows that the result is the same as the embedding we computed previously by averaging the two word embedding vectors.

But the convenience of using `DocumentPoolEmbeddings` comes at a cost – they often don't deliver as good performance as the following embedding methods.

DocumentRNNEmbeddings

`DocumentRNNEmbeddings` is a special type of document embeddings, implemented in Flair, that uses recurrent neural networks to form document vector representations.

It usually yields better performance than `DocumentPoolEmbeddings` and, in the same manner, receives word embedding objects as input. However, out-of-the-box `DocumentRNNEmbeddings` is useless – a newly initialized embedding object is essentially an untrained neural network model with randomized weights. The model needs to be trained in order to be useful.

But instead of getting sidetracked by training yet another type of embedding, let's instead explore a different kind of document embedding in Flair that is both pre-trained and yields excellent performance.

TransformerDocumentEmbeddings

A good approach to solving any problem is not to reinvent the wheel but instead to make use of technologies that already work well. A special kind of NLP model that has seen immense progress in recent years is transformers. Flair offers a wrapper class that is capable of computing embeddings for a whole document directly from a pre-trained transformer.

`TransformerDocumentEmbeddings` is used the same way as any other embedding class is used in Flair – it can be used to embed sentences using the `.embed(sentence)` method. Let's try this embedding out using the `roberta-base` transformer model:

```
from flair.embeddings import TransformerDocumentEmbeddings

embedding = TransformerDocumentEmbeddings('roberta-base')
sentence = Sentence('two words')

embedding.embed(sentence)

print(sentence.embedding)
```

The preceding code snippet will print out a 768-dimensional vector representation of our sentence. The class uses **Hugging Face** transformers and thus allows us to use a wide variety of models, which are fetched directly from the publicly available Hugging Face Models Hub. We will cover a lot more about Hugging Face and learn how to publish our own models on the Models Hub in *Chapter 10, Hands-On Exercise – Building a Trading Bot with Flair*.

There is a number of transformer-based pre-trained embedding models that are used by the `TransformerDocumentEmbeddings` class. The most notable examples are the following:

- `bert-base-uncased`
- `roberta-base`
- `distilbert-base-uncased`
- `xlnet-base-cased`

We have now covered an essential underlying NLP concept that fuels Flair text classifiers. We're now ready to dive deeper and explore how Flair text classifiers are designed.

Text classifiers in Flair

Most text classifiers in Flair utilize the `TextClassifier` class. It is responsible for providing the tools for defining and using text classifiers in Flair. Unlike many other concepts in Flair, it is a surprisingly simple design. `TextClassifier` is a model that utilizes a single linear neural network layer, taking the document embeddings as input and returning the class labels as output.

Let's learn how to train text classifiers in Flair as part of a hands-on exercise. Let's build our own sentiment analysis text classifier!

Training a text classifier in Flair

In this section, we will be training a sentiment analysis text classification model capable of labeling text as *positive* or *negative*. Text classifier training follows a sequence of steps very similar to how sequence labeling models are trained.

The steps required to train text classifiers in Flair include the following:

1. Loading a tagged corpus and computing the label dictionary map
2. Loading and preparing the document embeddings
3. Initializing the `TextClassifier` class
4. Training the model

The process, given what we covered as part of sequence labeling model training, should look very familiar – and indeed it is. Let's start by loading the data required to train the classifier.

Loading a tagged corpus

Training text classification models requires a set of text documents (typically, sentences or paragraphs) where each document is associated with one or more classification labels. To train our sentiment analysis text classification model, we will be using the famous **Internet Movie Database** (**IMDb**) dataset, which contains 50,000 movie reviews from IMDB, where each review is labeled as either positive or negative. References to this dataset are already baked into Flair, so loading the dataset couldn't be easier:

```
from flair.data import Corpus
from flair.datasets import IMDB
import _locale

# workaround for a rare Windows-only Flair encoding bug
_locale._getdefaultlocale = lambda *_: ['en_US', 'utf8']

corpus = IMDB()
corpus.downsample(0.05)
lbl_type = 'sentiment'
label_dict = corpus.make_label_dictionary(label_type=lbl_type)
```

The preceding code snippet loads the IMDB dataset into memory. It also creates the `label_dict` object, which is a dictionary containing all possible class labels. This will be required later when we build the classifier.

> **Important Note**
>
> In this code example, we utilized the `downsample()` method. The method downsamples the dataset. It reduces the number of items in our dataset, and by doing that, we also reduce the amount of data our model will be able to learn from. As a result, the model is likely to perform worse. This method should, unless there really is too much data, only be used for learning and demonstration purposes. We used a downsample rate of `0.05`, meaning we only keep 5% of the original data.

Let's now move on to loading the document embeddings into memory.

Loading and preparing the document embeddings

In the previous section, we covered all the different types of document embeddings that we can use in Flair. To keep this exercise simple, we will use `DocumentPoolEmbeddings`. They require no training prior to training the classification model itself:

```
from flair.embeddings import (DocumentPoolEmbeddings,
                              WordEmbeddings)


glove = WordEmbeddings('glove')
document_embeddings = DocumentPoolEmbeddings([glove])
```

In the preceding code snippet, we used **GloVe** embeddings. We passed them to the `DocumentPoolEmbeddings` class, which, if you remember the previous section well, simply computes an average of all the word embeddings in the document.

With document embeddings out of the way, we can now define our text classification model.

Initializing the TextClassifier class

This is the point where all the variables defined in the previous sections come together as part of a newly initialized object – `TextClassifier`:

```
from flair.models import TextClassifier


classifier = TextClassifier(document_embeddings,
                            label_dictionary=label_dict,
                            label_type=lbl_type)
```

In the preceding code snippet, we define our text classifier object, which initializes as a fully functioning classifier with randomized weights. Classifiers such as `TextClassifier` need to know the total number of classes in advance, as it affects the neural network architecture – hence the need to provide `label_dict`. We also need to pass in the document embeddings class at the point of defining the model so that the model knows the embedding length, which defines the number of input nodes. Finally, we also pass in the `lbl_type` variable, which is used at a later point.

Our model is actually a fully functional text classification model now. The tiny problem is that its weights are randomly initialized, which means that the output it produces will be useless. To make our model useful, we need to train it.

Training the model

Training the text classifier model involves two simple steps:

1. Defining the model trainer class by passing in the classifier model and the corpus

2. Setting off the training process passing in the required training hyperparameters

The following classification model training syntax is very similar to how we trained sequence labeling models in the previous chapters:

```
from flair.trainers import import ModelTrainer

trainer = ModelTrainer(classifier, corpus)

trainer.train('classifier',
              learning_rate=0.1,
              mini_batch_size=32,
              max_epochs=40)
```

The preceding code snippet sets off the training process where we pass in a number of training-related hyperparameters, all of which were explained in detail in *Chapter 5, Training Sequence Labeling Models*. The training process will now run for 40 epochs or until the model stops improving and will store the final model in the `classifier` directory.

> **Important Note**
>
> In the preceding exercise, we used the `DocumentPoolEmbeddings` document embeddings. However, if you ever plan to use `TransformerDocumentEmbeddings` (which is almost guaranteed to offer better performance), you should use the `ModelTrainer.fine_tune()` method instead of `ModelTrainer.train()`, which presets some of the training hyperparameters to aid fine-tuning transformers.

Training should take about 30 minutes on an average CPU. When done, we'll be ready to test the new classifier out.

Upon completion, the model trainer will print out a bunch of metrics. The simplest and most telling metric to look out for is the `test_score` metric. It shows the classification accuracy of the test dataset. In our case, `test_score` was `0.672`, although be aware that yours will be slightly different due to the randomly initialized weights.

Loading and using custom text classifiers

After training the text classification model, the resulting classifier will already be stored in memory as part of the `classifier` variable. It is possible, however, that your Python session exited after training. If so, you'll need to load the model into memory with the following:

```
from flair.models import TextClassifier

classifier = TextClassifier.load('classifier/best-model.pt')
```

We can now generate predictions on some made-up text input:

```
from flair.data import Sentence

sentence = Sentence("great")
classifier.predict(sentence)

print(sentence.labels)
```

The preceding code snippet tests our sentiment analysis classifier on the most trivial problem we could think of. We are classifying a single word, `great`, which, we can all agree, is unambiguously positive. The code snippet will print out the predicted class along with the prediction probability:

```
['Sentence: "great"'/'POSITIVE' (0.9999)]
```

This indicates that the model is 99.9% positive that the label is… well, `POSTIIVE`.

But be aware that the model was initialized on random weights and trained on a significantly downsampled dataset. The results will, therefore, likely be a little different every time you run this code.

We can give another simple example a go using the exact opposite. We can test whether our classifier behaves as expected when given an unambiguously negative message:

```
sentence = Sentence("bad")
classifier.predict(sentence)

print(sentence.labels)
```

You don't have to be an expert in sentiment analysis to determine that the word bad should be labeled as NEGATIVE, but neither does our model. It prints out the following:

```
['Sentence: "bad"'/'NEGATIVE' (0.9998)]
```

This shows that the model is 99.9% certain the word bad is negative, although this shouldn't be particularly shocking.

We should test our model on a slightly more complex example:

```
sentence = Sentence("Not quite my cup of tea")
classifier.predict(sentence)

print(sentence.labels)
```

The preceding code example presents a much more complex sentiment analysis problem, as it tries to predict a label for a document that doesn't necessarily contain any words that are positive or negative on their own.

The prediction we get for the preceding sentence is as follows:

```
['Sentence: "Not quite my cup of tea"'/'POSITIVE' (0.7513)]
```

Obviously incorrect. The preceding sentence clearly had a negative connotation. It's likely that when you run the code yourself and your weights are initialized differently, the outcome will be different, but it's certain that the prediction confidence (in our case, 75%) will be consistently low.

There are several steps we can take to improve the prediction quality of our classifier. The most obvious choice may be to simply downsample our corpus less and spend more time on training. It would be a valid choice, but it would be unlikely to help us improve the prediction quality by a huge margin. The biggest bottleneck to improving the classification accuracy, in this case, is the choice of document embeddings. We used DocumentPoolEmbeddings in our exercise. It is quick and simple to use, but it can only do so much. When training text classification models with the aim of getting good results, you should use TransformerDocumentEmbeddings instead.

In the preceding text classification exercise, we used a method called `downsampling` to reduce the size of our corpus, which significantly sped up the training process, but it also significantly affected our prediction quality. To improve the classification quality, we can simply downsample our corpus less. Originally, we threw out 95% of the dataset. In an effort to get better prediction quality, we could throw away only 80%, maybe only 50%, or maybe not throw away any data at all. But at this point, we'd hit a brick wall. We wouldn't be able to improve our classifier by training it on more data because there simply wouldn't be more data. In our case, as part of the exercise, we were able to leverage work of hundreds of man-hours of people hand-labeling the entire dataset entry one by one. But in the real world, the problem you'll be trying to solve most likely won't have a labeled dataset waiting for you to take advantage of. To make matters worse, labeling data is expensive. So much so that sometimes, hand-labeling data isn't even worth the effort, and the resources required to build a model will outweigh the benefits.

But what if there was a way to train a classifier without having to rely on many or possibly any training examples? Actually, there is a way. It's called **Task-Aware Representation of Sentences** (**TARS**). Let's learn how to use it for Flair.

Working with text classifiers that require little to no training data

When mentioning Flair and its strengths in the previous chapters, we mainly focused on various powerful ways of solving sequence labeling tasks. When talking about text classification, however, Flair was generally presented as a decent text classification tool, although nothing special compared to its sequence labeling capabilities. This was indeed the case – until now. Flair recently introduced a novel text classification method called TARS. The concept is described in depth in the *Task-Aware Representation of Sentences for Generic Text Classification* paper available at `https://aclanthology.org/2020.coling-main.285/`, which is well worth a read.

Transformer-based text classifiers in Flair leverage a special linear layer on top of the transformer model to produce the class probability distributions. The first problem with this approach is that when new class labels are introduced to the problem, or when we try to use different class labels (something that happens often in real-world machine learning), we basically need to discard this linear layer and train from scratch. The second problem with this approach is that the model learns about the meaning of the classes purely based on the connection between the training examples and the hand-labeled class labels. It doesn't extract any knowledge from the semantic meaning behind the class label word. For example, in the exercise from the previous section, we performed sentiment analysis with the IMDB dataset using the `NEGATIVE` and `POSITIVE` class labels. With classical text classification approaches, the class labels are simply unique identifiers of the classes and hold no semantic meaning. If we tweaked the labeled corpus and renamed NEGATIVE to CLASS_1 and POSITIVE to CLASS_2, the results would be exactly the same. TARS overcomes both of these problems with a novel text classification technique that uses the transformer model itself and leverages the class label word semantic information from the transformer.

This allows TARS to perform text classification with the help of only a few training examples (few-shot classification) and, sometimes, even no training examples at all (zero-shot classification). TARS outperforms previous approaches on small training data. Best of all, it's fully implemented and ready to use Flair.

Let's give both few-shot and zero-shot text classification with TARS a try!

Zero-shot classification with TARS

Zero-shot classification with TARS leverages the semantic information of the class label words captured inside a pre-trained transformer without the use of any training data.

Using TARS for zero-shot text classification in Flair couldn't be simpler. You need to follow the following three steps:

1. Load the pre-trained TARS model for the right language.
2. Define the classes (which the TARS transformer model hopefully knows about).
3. Use the predict_zero_shot() method to generate predictions.

The preceding steps may sound like a lot of work or at least something that requires some level of writing code. Instead, each step is merely a one-liner.

Load the pre-trained TARS model for the right language

Before getting our hopes of not needing any training data up, we need to first make sure there is a model available for us to use for the right language. In Flair, as of version 0.11, there is only a single model available. It is an English model and is called **tars-base**.

Let's see how tars-base's zero-shot text classification handles sentiment analysis.

We load a TARS model by calling the TARSClassifier.load() function:

```
from flair.models import TARSClassifier

tars = TARSClassifier.load('tars-base')
```

This will download the TARS model and store it in memory, as well as also storing it to disk for future use. We can now define the class labels.

Defining the classes

Class labels for TARS models are simply a list of strings. We can therefore define our sentiment analysis classes as the following:

```
classes = ["positive", "negative"]
```

And that's all there really is to it – no training or training examples required! Let's generate some predictions.

Generating zero-shot predictions with TARS

Let's test our TARS model using the same input we used to test our DocumentPoolEmbeddings based classifier that leveraged GloVe embeddings. We'll see whether tars-base can also classify this single-word sentence, great, correctly:

```
from flair.data import Sentence


sentence = Sentence("great")
tars.predict_zero_shot(sentence, classes)


print(sentence)
```

We generate the predictions using the predict_zero_shot() method. The preceding code snippet will print out the following:

```
Sentence: "great" → positive (0.9691)
```

This indicates that the predicted label is positive with 96.9% confidence.

That's pretty impressive given zero training data! Let's see whether the same successful outcome is achieved using a single-word sentence with a negative connotation:

```
sentence = Sentence("bad")
tars.predict_zero_shot(sentence, classes)


print(sentence)
```

This prints out the following:

```
Sentence: "bad" → negative (0.9037)
```

This clearly indicates that the word bad is associated with a negative sentiment – or rather with the semantics of the class label word negative.

Finally, we can test our zero-shot classifier on the Not quite my cup of tea sentence that our previous text classifier wasn't really sure about:

```
sentence = Sentence("Not quite my cup of tea")
tars.predict_zero_shot(sentence, classes)


print(sentence)
```

This prints out the following:

```
Sentence: "Not quite my cup of tea" → negative (0.9859)
```

Hurray! We have a clear winner. tars-base clearly, with 98.6% confidence, predicts the sentence Not quite my cup of tea as negative – or rather associates the sentence with the semantic meaning of the word negative. Of course, to determine that our classifier indeed performs better than our basic DocumentPoolEmbeddings classifier, we'd need to perform some means of evaluation. But instead of doing that, let's explore how we can further improve our classification model using few-shot learning.

Few-shot classification with TARS

TARS models allow us to use only a handful of labeled examples to train a text classifier callable of delivering excellent performance.

Let's explore TARS few-shot learning through solving the same problem that we tackled in the previous section – sentiment analysis on the IMDB dataset.

Training a TARS model is hardly any different from what we did with training DocumentPoolEmbeddings classifiers. It involves the following steps:

1. Loading a tagged corpus and computing the label dictionary map
2. Loading a TARS model and setting the task
3. Training the model

Let's start with loading the dataset. But bear in mind we'll need much, much less data than one would expect.

Loading a tagged corpus

We'll be loading the IMDB corpus the same way we did in the previous section. But this time, we'll be downsampling it much more significantly. After all, we are doing few-shot learning. And if the claims are true, we should be able to get excellent performance using only a few training examples.

Let's load the dataset and downsample it to 1% of its original size:

```python
from flair.data import Corpus
from flair.datasets import IMDB
import _locale


# workaround for a rare Windows-only Flair encoding bug
_locale._getdefaultlocale = lambda *_: ['en_US', 'utf8']


corpus = IMDB()
corpus.downsample(0.01)
lbl_type = 'sentiment'
label_dict = corpus.make_label_dictionary(label_type=lbl_type)
```

With the preceding code snippet, we loaded the IMDB dataset that originally contained 50,000 movie reviews from IMDB. Each review is labeled with either a POSITIVE or a NEGATIVE label. We then downsample the corpus to 1% of its original size – to only 500 reviews. We then compute the label dictionary, which will come in handy soon.

Loading a TARS model and setting the task

We will now load a TARS model the same way we did in the zero-shot learning example. But this time, we'll need to tell the model that it'll be retrained with the aim of predicting two specific labels. We will do this by using the add_and_switch_to_new_task() method:

```python
from flair.models import TARSClassifier
from flair.data import Sentence


tars = TARSClassifier().load('tars-base')
```

```
tars.add_and_switch_to_new_task(
    task_name="sentiment_analysis",
    label_dictionary=label_dict,
    label_type=lbl_type)
```

In the same manner as before, we load the `tars-base` model. But this time, we pass in some extra info, telling the model that it'll be performing a specific task. It'll be classifying text into classes defined as part of the `label_dict` variable.

We are now ready to train the actual model.

Training few-shot TARS models

Training few-shot models in and of itself isn't any more complex than training any other model in Flair. The only significant difference is the amount of compute power required. In most prior exercises in this book, you could get away with doing model training on a CPU where the only sacrifice was slightly lower model performance due to fewer epochs or smaller model size. But here, we are dealing with a fully sized transformer model, and there isn't much we can do to reduce the needed compute power.

You *will* need a GPU-equipped machine or a **Google Colab** GPU runtime to successfully run the following code.

The following training syntax should, at this point in the book, look very familiar:

```
from flair.trainers import import ModelTrainer


trainer = ModelTrainer(tars, corpus)
trainer.train(base_path='few-shot-tars',
                learning_rate=0.02,
                mini_batch_size=16,
                mini_batch_chunk_size=4,
                max_epochs=6)
```

In the preceding code snippet, we instructed the model trainer to train our TARS model for six epochs (when training production models, this number should be set higher). We also set batch size and batch chunk size to a very low value to increase the chances of our machine being able to handle it.

This will now start training, which will last about 30 minutes on an average GPU instance. Once done, the model will be stored in `few-shot-tars/best-model.pt`, relative to where you ran the code from. In Google Colab, you will be able to access the model without a problem as part of the same runtime by pointing the code to its location at `few-shot-tars/best-model.pt`. However, the model will disappear if you restart the Colab notebook runtime.

Using custom few-shot TARS models

When training finishes, the trained model will be stored in memory. But if, for whatever reason, your Python session died, you can load the model from disk instead by running the following:

```
tars = TARSClassifier().load('few-shot-tars/best-model.pt')
```

Good. We can now generate some predictions. Let's use the model and give it the same exact input as our previously trained models and observe whether there's any improvement.

Let's start with the most basic example. Let's predict with the word `great`:

```
from flair.data import Sentence

sentence = Sentence("great")
tars.predict_zero_shot(sentence, classes)

print(sentence)
```

This prints out the following:

```
Sentence: "great" → positive (0.9991)
```

A 99.9% certainty! This is an apparent improvement over the 96.9% we got with the zero-shot model.

Let's try another trivial problem – the word `bad`:

```
sentence = Sentence("bad")
tars.predict(sentence)

print(sentence)
```

This prints out the following:

```
Sentence: "bad" → NEGATIVE (0.946)
```

Nice! That's another improvement over the zero-shot model. Now, let's try our slightly more complex sentence:

```
sentence = Sentence("Not quite my cup of tea")
tars.predict(sentence)

print(sentence)
```

This prints out the following:

```
Sentence: "Not quite my cup of tea" → NEGATIVE (0.9921)
```

That's the best result so far!

Important Note

In our exercise, we often assess model performance by comparing results based on only a few examples. This is simply to illustrate how our new models perform in general. To get a conclusive answer on which model performs best, you should always use proper evaluation techniques and evaluate the models on the same data.

Our TARS few-shot model achieved comparable or better performance than our previous models while trained on only 1% of the original dataset for only six epochs. Truly remarkable!

With this remarkable state-of-the-art NLP technique, we conclude the topic of text classification.

Summary

This chapter provided an overview of the text classification methods in Flair. We first went through some basics and theory that set the stage for the hands-on text classification tasks. We began with exploring the classic text classification techniques and trained our very own model on the IMDB dataset. Later, we learned about TARS classifiers and witnessed their amazing few-shot and zero-shot classification capabilities.

This not only concludes the text classification section of this book; it wraps up the entire part of the book that explores what Flair can do. We're now armed with knowledge and experience in sequence tagging, word embeddings, Flair's set of pre-trained models, training our own sequence tagging models, hyperparameter tuning, and much, much more.

There's nothing stopping us from putting some code together and deploying a real-world NLP model in the wild… well, nothing apart from not being quite sure how to deploy, host, or maintain these models as part of a production-grade service. We'll be learning about that in the upcoming chapter.

Part 3: Real-World Applications with Flair

In this part, you will learn how to deploy and use Flair models in production. You will go through a set of practical exercises that use all the knowledge you have acquired throughout this book to solve real-world problems.

This part comprises the following chapters:

- *Chapter 9, Deploying and Using Models in Production*
- *Chapter 10, Hands-On Exercise – Building a Trading Bot with Flair*

9
Deploying and Using Models in Production

When **Natural Language Processing** (**NLP**) researchers and scientists work on improving a certain NLP model, they usually spend significant amount of time working on improving some very specific part or aspect of the model. When their work is done, results are gathered and then published as part of an academic paper. However, if you plan to leverage NLP for practical purposes, either as a commercial solution or simply as part of a personal project, a well-trained model is usually the point where your journey only begins.

In the first section of this chapter, we plan to cover the typical issues generally encountered when using NLP models in production, how to overcome these issues, and how to make our models publicly available. We will then address the issue from the low level by building a self-serving NLP solution. We will talk about the tools and libraries available for self-serving and the reasons why you should or shouldn't opt in for self-serving. Finally, we will cover the various types of managed NLP machine-learning-as-a-service solutions, and how and when to use them. We will wrap up with a hands-on exercise where we deploy Flair models to production using the **Hugging Face Accelerated Inference API**.

In this chapter, we will talk about how to tackle these problems and deploy our own Flair models to production with ease as part of the following sections:

- Technical considerations for NLP models in production
- Self-serving Flair models
- Using managed services for deploying and using Flair models

Technical requirements

To run the code in this chapter, you will need the usual local Python setup with Flair version 0.11. You will also need one additional Python library called **Flask** version *2.0.3*. Code examples covered in this chapter are found in this book's official GitHub repository in the following Jupyter notebook: `https://github.com/PacktPublishing/Natural-Language-Processing-with-Flair/tree/main/Chapter09`.

Technical considerations for NLP models in production

Deploying machine learning (especially NLP) models differs from deploying other software solutions in one key area – the resources needed to run the service. Hosting a generic service for a simple mobile or web app can, in theory, be done by any modern device such as a PC or a mobile phone. Only as you start to scale the service to cater to a larger audience is when you need to put extra thought, effort, and resources into making the service more scalable. When serving machine learning models, things often get complicated right from the start. We are dealing with huge models where a typical web server sometimes can't even load a model into memory. Each request uses up a significant amount of resources, yet we need to serve requests on demand, in real time, and to a large audience. But how do you do that?

When comparing this chapter's topic to what we covered in this book so far, you will notice that deploying models to production is an entirely new problem that requires a slightly different set of skills. What we did in the preceding chapters was focused on the understanding and processing of data and the utilization of different machine learning and natural language processing techniques to achieve a certain goal. All this knowledge is less relevant now. What we care about now is deploying and maintaining NLP services in a way that will allow them to serve as many requests as quickly as possible, as reliably as possible, and using as few resources as possible. Easy, right?

The set of skills to achieve this is often so different from the set of skills required to design and train NLP models that it's typically handed over to completely different individuals – the software or **MLOps** (a field joining machine learning, DevOps and data engineering) engineers. There is a great book published by Packt called *Engineering MLOps* where you can learn more about this interesting engineering field. But while deploying ML models to production requires a very specific set of skills, it's not an impossible process in and of itself and can be done by anyone who follows these steps.

Before you make any decisions about deploying your NLP model, you need to first ask yourself the following questions:

- Is the inference (the process of generating predictions) going to be done client-side or server-side?
- Does the service need to be scalable?
- Do you want to learn by doing this work or simply achieve a goal?
- Is your model deployable on a managed service?

All these decisions play an important role in determining where and how the NLP models will be deployed. Let's investigate each point in detail.

Client-side versus server-side inference

Probably the most important and, at the same time, the easiest to answer question relating to deploying NLP models is whether they will be stored and perform inference locally on a client's machine, or whether the model will be stored and the inference performed remotely on a server or a managed cloud service.

When machine learning inference is performed locally on a client's machine (such as a mobile phone, laptop, PC, or an e-book reader), we call this client-side inference.

When machine learning inference is performed remotely on a server or a managed cloud service, we call this server-side inference.

Both approaches have their pros and cons, and deciding which one to use almost always depends on the problem you are trying to solve with the NLP model because, at this point, you already know what NLP problem the model solving; this means that the decision is generally very straightforward.

Because machine learning models are typically large and require significant resources, there is a general misconception in the ML community that all machine learning inference needs to be done server-side on large GPU-equipped machines.

Surely, given the recent growth in the size of transformers and other NLP models, the era of using machine learning models locally is over? My mobile phone or laptop could never handle production-grade machine learning models, right?

As plausible as the preceding sentence sounds, it is undeniably and utterly false. If you are using an electronic device to read this book, chances are that the machine you are using is packed with machine learning models.

Modern electronic devices use locally stored machine learning models to solve a number of different tasks:

- *Face recognition on mobile phones, laptops, and PCs*: Modern electronic devices use computer vision models optimized to run on low-performance devices to perform face recognition instantaneously without the need for a network connection. It would be rather unfortunate if you were only able to unlock your phone or laptop if your internet connection worked.

- *Fingerprint recognition on mobile phones*: Fingerprint recognition is yet another method of unlocking a device with emphasis on speed. A user expects their fingerprint to be recognized in a matter of milliseconds. Aside from speed, another major reason why storing a fingerprint recognition model locally makes sense, is privacy. Most users would be appalled if they realized their fingerprint data is being streamed to a server located on the other side of the planet owned by a big tech corporation, even if the company's intentions were completely non-malicious. Low latency and privacy are key for fingerprint recognition, which is why manufacturers almost always opt in for client-side inference.

- *Typing predictions for virtual keyboards*: Another area where low latency and privacy outweighs prediction quality is virtual keyboards. Software keyboards download and store typing prediction models locally, which allows users to leverage typing predictions even when offline in environments such as an underground train.

- *Self-driving cars*: Self-driving cars make use of a number of machine learning models that allow them to make decisions about navigating a vehicle.

- *AI camera apps on mobile phones*: Modern mobile phone camera applications leverage specialized machine learning algorithms that select the right camera settings based on the environment and then post-process the image using machine learning-based image processing software. The models need to be fast, responsive, and available consistently – a perfect match for local inference.

The list goes on, and it clearly shows that there currently is a demand for specialized low-resource machine learning models designed for local inference. It's entirely possible the trends may lean more towards the server side in the future, but it's unlikely local models will disappear completely. This is simply due to the physical laws that govern the theoretically minimal signal latency. For example, even in the theoretically best possible scenario where we develop a server-side machine learning service that brings inference time close to zero and build a network infrastructure that sends signals at the speed of light in a vacuum, the amount of time needed to send an inference request from a device in Paris to a server in New York and back will still take 54 milliseconds (the amount of time light travels from Paris to New York and back in a vacuum). So even the theoretically best possible scenario for server-side inference is still too slow for some applications such as self-driving cars.

Both server-side inference and client-side inference clearly come with a set of advantages and drawbacks, and before making a rushed decision, it's important to consider the pros and cons.

The advantages of using local inference:

- *Privacy*: To an average engineer who understands how safe network protocols and communication can be, the thought of a photo of their face or a scan of their fingerprint being transmitted over the internet isn't off-putting or scary. However, for the non-tech-savvy, this spells disaster. When the news of highly personal or biometric data being streamed over the internet draws media attention, the impact for the company will be significant. It doesn't matter whether the company developed the safest possible algorithm, one that could even be safer than local inference; the media and people appalled by this fact won't care, and the damage will be done. When dealing with personal and biometric data, it's therefore important to take into careful consideration how and where this data is going to be transmitted and stored. A good way to avoid this problem altogether is to not transmit the data at all and use local inference instead.

- *No network latency*: Powerful servers and supercomputers have the potential to do inference quicker than an average mobile phone, but if we decide to go down the remote inference route, our requests and results will need to travel over the internet, incurring network latency. We can completely eliminate network latency by not doing network communication at all – with local inference.

- *Maintenance*: Optimizing a model for low-resource devices is hard. But once you get it right, and build a model that works on all different types of devices and deploy it successfully, your job is done. With local inference, there's no service to take care of or network spikes to monitor – just sit back and relax.

- *Cost saving*: As harsh as it sounds, it's true. When a company does client-side inference instead of server-side inference, it means that all the computations required to generate results as part of the machine learning models will be done on customers' devices. These computations will use electricity. And the electricity is paid for by... the customer.

- *Scaling by design*: When we go from 1 user to 10 users with server-side inference, the amount of compute power required will increase tenfold. With local inference, it's different. Inference is distributed by default. We have a dedicated machine learning inference machine for each user – their own device. It doesn't matter whether a single user does inference on their machine or whether it's a billion users. This leads us to an interesting conclusion – *local inference delivers scaling by design.*

The disadvantages of using local inference:

- *Catering to resource and performance constraints*: The model you plan to deploy will likely need to work on a number of different devices with varying compute capabilities. Should you deploy a very simple dumbed-down model that'll work on all devices but won't deliver great performance? Should you deploy models on high-end devices only and simply tell the low-end device users they can't use the feature? Or should you build different models for each device type – some models for low-end devices and some for high-end devices? These are all questions you'll need to answer if you go down the local inference route, and the answers will depend on business needs and requirements.

- *Compromised prediction quality due to size and resource constraints*: Local inference models need to be small. They need to fit on a device such as a mobile phone and be stored in its very-limited memory. This means that the model will be hugely limited in terms of the number of parameters. To optimize a model for local inference, you may need to reduce it in size significantly. This undoubtedly results in a huge hit for prediction quality.

- *Versioning hell for model updates*: If you host a model on a remote service, to update a model with a newer, improved version, you simply need to swap the old model with the new one on the remote service. However, with local inference, things get complicated. To make model updates possible, you first need to implement a system that manages updates. You then need to release the new model to all the devices that have an internet connection and have opted in for updates, but you also need to keep in mind that there will be a certain proportion of users who will never update due to no internet connection or refusal to accept updates. As a result, your user base will be using a variety of different versions of models, and you will need to support them all.

Conversely, there's a long list of advantages to using server-side inference:

- *Server-side applications can make use of the out-of-the-box, machine-learning-as a-service solutions*: If you opt in for server-side inference, a whole new world of ready-to-be-used solutions awaits that likely requires a few clicks to get your model from a file to a production service. There is a long list of managed services you can make use of if you plan to do inference server-side, whereas for local inference, you will need to, at least to some extent, reinvent the wheel.

- *You can use state-of-the-art models*: Current state-of-the-art NLP models have a large number of parameters, and we can't expect any of them to fit on a device such as a mobile phone anytime soon. To use state-of-the-art NLP models in production, you simply need to resort to server-side inference.

- *Updates are easy and comprehensive*: To update a model server-side, unlike with local inference, you simply need to swap the old model with the new one at a single point. Once you do that, all inference performed from that point onward will feature results generated by the new model.

At the beginning of this section, we mentioned that deciding between client-side or server-side inference is usually a simple question to answer. All it really boils down to is whether we want huge high-performance models with network latency or no network latency at all but compromised model quality.

The server-side inference approach simplifies a number of aspects of NLP model deployment but also raises a number of new questions. Let's discuss these as part of the next section.

When to use managed services and when to self-serve server-side models

Another important decision that has a huge impact on how you manage your NLP services boils down to whether you will choose a managed service or decide to self-serve.

If we decide to self-serve our models, it'll mean that we'll be responsible for setting up server infrastructure, building software for the endpoint that delivers NLP results, building software for performing inference on the model, scaling the system when our user base grows, and much, much more. We will essentially be taking care of the entire machine learning life cycle manually.

Conversely, a managed or fully managed service is a commercial solution that provides support for most of these tasks. Fully managed machine learning services provide out-of-the-box support for the following:

- *A managed infrastructure*: When deploying NLP services, the final model will likely live on a particular machine somewhere in a data center. In order for our users to use our service, they will need to have means of reaching this machine through a domain name or an IP address. To set up a self-serve NLP service, infrastructure needs to be put in place to make sure that the right requests hit the right server so that it can return the results to the user. All these decisions are taken care of for us as part of a managed service.

- *Managed endpoints*: A machine learning model, from an engineering perspective, is a binary file that can be loaded into memory and, through a machine library, perform interface. However, given that we are using a server-side inference approach, it means that if a user needs some predictions from an NLP model, they will need to send a request over the internet. This request will need to reach the service, which will need to interpret the request and respond appropriately with results. Therefore, there needs to be a means of communication agreed by the client and the service. A popular solution to this is using a communication protocol called **REST**, which allows communication through a **REST API** that utilizes the **HTTP** or **HTTPS** protocol. But the machine learning endpoints are not just responsible for client-server communication, parsing requests, performing inference, and returning the results to the user. They have a number of other responsibilities. They need to return appropriate error messages if something goes wrong (for example, if the input text exceeded the length limit). They need to return appropriate error messages if the user doesn't have access. They need to be able to handle a large number of requests at the same time and queue appropriately. They need to perform optimization tasks such as caching.

When building a production-grade NLP service, you will need to think about these issues and implement solutions yourself. Managed services take care of all that.

But it's worth mentioning that managed services can only do so much. Self-serving, while cumbersome, allows you to implement whatever you want and add any feature you can think of. With managed services, you are limited to using only the features offered by the service.

- *Scaling*: One of the biggest selling points of fully managed ML services is the promise of scaling. It resolves one of the biggest issues with self-serving, which is getting caught up in a non-scalable infrastructure. For example, let's imagine you implement a simple ML endpoint on a single virtual machine. This solution will be easy to implement and work great for a few daily users. But when you start getting more users, the service will start returning delayed results, and too much traffic will sooner or later cause a denial of service. The application will then need to be modified to be able to handle a medium amount of daily users. By the time this solution is implemented, you will have lost a great deal of customers due to service unavailability. When a solution capable of dealing with a medium amount of traffic is implemented, you may realize that the number of users is increasing too quickly and that the entire solution will need to be reimplemented again to make it high-traffic capable.

  Fully managed ML services claim to solve all these problems. They advertise their services as scalable by default. This means that if your service works great for 1 daily user, it should, in theory, work the same for 1 million daily users. This allows you to spend time and resources on more useful endeavors.

- *Deploying models and versioning*: Most fully managed ML services come with a set of tools and features for deploying ML models. In some cases, using them is as easy as uploading the ML model binary and clicking **deploy**.

- *Monitoring*: If you decide to self-serve, you will inevitably also need to build or set up tools for monitoring. These tools provide insights about how many users are using the service and help you decide whether or not you should scale. Nearly all managed services offer this out of the box.

- *Freemium pricing plans and low setup costs*: Most fully managed ML services have a near-zero setup cost and a pricing plan where you pay only for what you use. Furthermore, some services offer a freemium service where the first N daily/monthly requests are free. Pricing for these services is often competitive and affordable compared to the spending resources with self-serving. It is only when a company grows to a significant size (such as becoming a big corporation) that a fully managed service will become less economical than hiring a dedicated team responsible for self-serving.

> **Important Note**
>
> Fully managed ML services also feature tools for the complete ML life cycle, which includes data labeling, data preprocessing, model training, deployment, and validation. These features are, by design, limited in terms of what they can do and often only support a limited set of NLP tasks, such as text classification, but they may not work for tasks such as sequence tagging or specifically training Flair models.
>
> Because few popular managed ML services offer support for the entire ML life cycle of all the types of models that Flair can build, we plan to focus only on the steps required to get your NLP model deployed to production. The steps required for setting up automation pipelines for maintenance and model refresh will be left out, as they relate to a more general field called MLOps.

The preceding bullet list shows a number of points clearly working in favor of fully managed services. This begs the question, *why not use a managed service over self-serving every time?*

The general answer to the preceding question is: if you can, you should. There are, however, some cases where self-serving may be a more reasonable choice. You can decide to self-serve if the following applies:

- You want to learn about how setting up ML services works from a low-level engineering perspective.
- You have a large enough user base that makes using managed ML services uneconomical.
- The ML library, framework, or task you are working on isn't supported by the managed service.

Unless you fall under any of the categories described in the preceding bullet list, you are likely to be better off simply using a fully managed service.

With this, let's have a look at what services you can use to deploy Flair models.

Managed services for Flair models

Managed services for hosting ML models are usually commercial solutions offered by either big-tech cloud platforms or specialized ML-oriented companies that offer ML as a service.

The list of services that can be used with Flair, however, is constantly changing, and so the user interfaces for interacting and managing these services. This means that a step-by-step guide focusing on how to interact with the UI or API of a specific managed ML service is likely to become outdated within a year.

However, you're not out of luck! There *is* one exception. There is one specific service that is tightly integrated with Flair and is likely to be supported for years to come. It's called the **Hugging Face Models Hub**. It's a repository that hosts a vast number of ML models, as well as providing tools to test them in a web browser, viewing instructions on how to train them, and deploying them as a service.

Later on, we'll have a look at how to train, host, and use Flair models deployed as part of the Hugging Face Accelerated Inference API service.

But before wrapping this section up, let's recap all the advice, tips, and tricks covered in this section.

A model deployment decision-making flowchart

To make Flair model deployment decision-making easier, we prepared a flowchart that helps you decide on the right path to deploying Flair models to production.

Depending on your business needs, resources, and aspirations, you will have these three options for deploying your Flair model:

- Client-side deployment with local inference
- Fully managed ML as a service solution
- Self-serving

The following flowchart will make it easier for you to decide between the preceding three approaches:

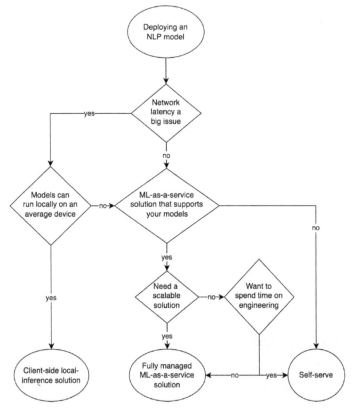

Figure 9.1 – The NLP model deployment decision-making flowchart

Depending on what situation you are in, you will end up using one of the three ML deployment types. Bear in mind that the proposed decision-making process is based purely on good practices and subjective advice. Everyone's situation will be different, and you should always use your own or other experts' judgment.

Regardless of which ML deployment route you choose to go down, it's always useful to have at least some level of understanding of each possible option.

Let's explore both self-serving and using managed ML solutions as part of the following hands-on exercises.

Self-serving flair models

Regardless of whether you decide to self-serve or use a fully managed service, having some basic understanding of how NLP model deployment inference works in production is always useful.

Since deploying NLP models in production is more of a software engineering feat than an NLP challenge, general software engineering rules apply – the main one being, *don't reinvent the wheel.*

While implementing your own NLP model-serving frameworks is entirely possible, doing so will require a significant amount of time. Chances are that even if you put your best efforts into implementing a solution, you won't be able to build a product better than the open source solutions out there built by hundreds of different contributors.

There is a wide range of tools and packages out there for self-serving NLP models, but one specific package stands out due to its support for **PyTorch** packages (Flair is built on top of PyTorch) and its support for Hugging Face models. The package is called **TorchServe**.

TorchServe for self-serving Flair models

PyTorch is a popular open source ML framework. Flair, an NLP framework on its own, is built on top of PyTorch and utilizes its tools for designing, training, and storing neural models.

Because Flair models are essentially PyTorch models, a tool such as TorchServe seems like a perfect fit. Let's see how TorchServe works by having a look at its design chart:

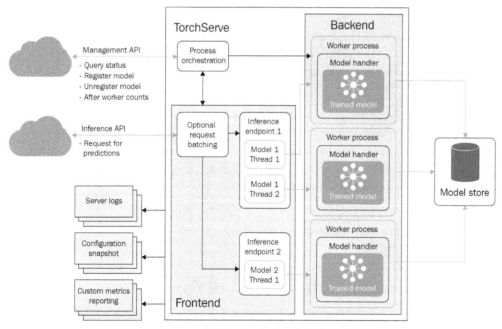

Figure 9.2 – TorchServe design

The preceding design schematic shows various parts of the TorchServe framework as described in their official documentation at `https://github.com/pytorch/serve`.

Let's take a look at the main components:

- **Frontend**: The component responsible for handling requests. It receives the user's request as input, parses it, and passes it on to the backend that returns the model output, which is then processed, formatted, and sent back to the user.

- **Inference endpoint**: A specific part of the frontend, reachable at a certain address, and responsible for handling requests and returning results.

- **Inference API**: An API that allows and defines communication between the client and the frontend.

- **Model store**: A directory that contains the models.

- **Backend**: A threaded component that contains model handlers responsible for performing inference.

- **Management API**: A special API, dedicated to the owners of the application, allowing us to configure the service.

The flow of events of a single use of a TorchServe service is as follows. A client first sends a request to the frontend using the inference API. The request contains a payload containing the input that will be provided to the NLP model. If the framework is too busy to process the request, the request is batched (queued). If not, it is passed to the endpoint, where it is parsed, and the payload is passed on to a model handler. A model handler performs inference on the loaded NLP model and returns the output to the endpoint. The endpoint prepares and formats the output (typically, using the JSON format) and returns it to the user.

The preceding TorchServe design run-through provides a very high-level overview of how it works and how it is used to serve NLP models. The setup and installation process of TorchServe can be very platform-specific and not something that can easily be covered in a few lines of code. The entire process of setting up TorchServe is described brilliantly as part of their official documentation on GitHub.

But don't let the fact that this book won't be covering TorchServe in great depths get you down. As part of the upcoming section, we'll implement our very own proof-of-concept tool for serving Flair models.

A proof-of-concept self-serve tool for Flair

Implementing a reliable, scalable, safe, and production-grade NLP self-serving tool is a complex process that requires weeks, if not months, of careful thought and preparation. In most cases, instead of implementing the solution yourself, you'll be better off using an already available open source tool such as TorchServe.

But sometimes, we simply want to build a proof of concept. This is a simple tool that demonstrates what our model can do and allows the service to be available to at least a handful of users/customers.

To learn about that, let's build a simple service that performs named entity recognition using Flair's state-of-the-art English **Named Entity Region** (**NER**) tagger.

The solution can be implemented by following these steps:

1. Installing the required libraries

2. Defining the function for sequence labeling

3. Defining the endpoint

4. Starting the service

Let's implement the solution step by step.

Installing required libraries

This proof-of-concept tool, aside from Flair, requires one extra library that isn't preinstalled as part of the Python standard library; neither is it a Flair transitive dependency (a requirement of a required library). This library is called **Flask**. It's a lightweight Python framework for setting up endpoints and is a perfect prototyping tool.

We'll install version 2.0.3 by running the following:

```
pip install flask==2.0.3
```

This will install Flask version 2.0.3, ensuring that your code behaves consistently.

Defining the sequence labeling function

Next, we need to define the function that performs named entity recognition on some input text. We'll be using Flair's default, state-of-the-art NER tagging model, which uses the ner model key:

```
from flair.data import Sentence
from flair.models import SequenceTagger


tagger = SequenceTagger.load('ner')


def tag_text(input_text):
    sentence = Sentence(input_text)
```

```
tagger.predict(sentence)

return sentence.to_tagged_string()
```

In the preceding code snippet, we define the `tag_text(input_text)` function using the standard sequence tagging syntax used many times across this book. The function receives text as input and returns tagged text as output. For example, for the input text `in New York`, it'll return the tagged version of the input string – `"in New York"` → `["New York"/LOC]`.

The preceding code revolves around how we define the `tagger` object. It's defined globally outside the tagging function, which ensures that the model is only loaded once and can be reused for any number of requests. Do note that this approach will cause issues for production-grade solutions where multiple processes attempt to use the same Flair model object.

We're now ready to define our endpoint.

Defining the endpoint

This endpoint is responsible for being available to receive requests at a certain address, parsing requests, passing them onto the `tag_text()` function, preparing the results, and passing them back to the user:

```
from flask import Flask, jsonify, request

app = Flask(__name__)
app.config['JSON_AS_ASCII'] = False

@app.route('/ner-service')
def query_example():
    text = request.args.get('input', '')

    tagged_text = tag_text(text).replace('"', '\'')

    return jsonify({"result": tagged_text})
```

The preceding code snippet starts by defining the `app` object for the web service. The following function is a definition of a web endpoint. We define the endpoint by prepending the function definition with the `@app.route('/ner-service')` decorator. This ensures that the endpoint is available at the `/ner-service` address.

We then fetch the input text straight from the URL as part of the `input` query string parameter. This means that if the user sends a GET request to `/ner-service?input=in+New+York`, the value of the `text` variable will be *in New York*.

> **Important Note**
> Using GET query string parameters as a means to receive text input from clients is almost always a bad idea. We used GET in this code example because it's easy to do debugging using only a web browser. For production-grade services, you should use POST instead.

The endpoint passes the input text to our `tag_text()` function and returns the result in **JSON** format, which allows the result to be both easily parsable by software as well as readable by humans.

And that's it! Now, we only need to start the service.

Starting the service

The service can then be started by running the following:

```
app.run(debug=False, port=9002)
```

The code will start a web service using port `9002` and will register a previously defined endpoint at the `/ner-service` address. Note that the preceding command will run forever or at least until our web service crashes. If you ever need to stop the service, simply terminate the execution of the code snippet.

We can now test our amazing NER tagging web service by directing the browser to our own machine's local IP address at `127.0.0.1`.

Let's first visit our endpoint without providing any input text by directing our web browser to `http://127.0.0.1:9002/ner-service`.

Our service returns the following:

```
{"result":""}
```

This indicates that the returned tagged text is an empty string. This makes sense given no provided input. Now, let's test it out with some actual input by visiting `http://127.0.0.1:9002/ner-service?input=Never+been+to+Lima`.

We expect the response to the preceding request to be a string where the word `Lima` is tagged as a location. Let's see:

```
{"result":"Sentence: 'Never been to Lima' → ['Lima'/LOC]"}
```

Hurray! It looks like our NER tagger service works as expected. You can now start this service on a web server or make your own development machine's IP address public – and you will make the service available to the world.

But before you do just that, keep in mind that our proof-of-concept solution has a long list of drawbacks and issues that will arise if the service is used for commercial purposes. The drawbacks include: no input sanitization, no request authorization, no caching, no multithread availability or thread safety in place, hard-to-parse output, no logging or monitoring… the list goes on. It clearly shows that such code should only be used for prototyping. For production-grade code, it almost always makes sense to use out-of-the-box solutions such as TorchServe.

Better still, you can avoid writing code altogether – by using a managed service. Let's learn how to do that in the upcoming section.

Using managed services for deploying and using Flair models

Unlike self-serving, where most aspects of the ML life cycle need to be taken care of manually, the managed and fully managed ML services sell the idea of a complete out-of-the box ML as a service solution.

Most of these services offer guarantees about service availability (what percentage of the time the service is guaranteed to be working) and scalability (the ability to scale without having to refactor the entire infrastructure every time the user base grows). Some services also offer management of the entire ML life cycle called **Machine Learning Model Operationalization** (**MLOps**) management. But some managed services may have trouble providing support for all the features and tasks Flair is capable of solving. This applies to almost all popular ML-as-a-service solutions, with one exception – the Hugging Face Models Hub.

The Hugging Face Models Hub

Hugging Face is an NLP-oriented company with a big open source community. The company became largely popular after releasing their transformers library (more information can be found at `https://huggingface.co/`). They also feature the Hugging Face Models Hub, which is a collection of NLP models and datasets, most of which are public, meaning they are free to download and can be used by anyone. Furthermore, models hosted on the Hugging Face Models Hub can make use of the Accelerated Inference API. This is a managed NLP service with a freemium pricing plan. This service is available for free at the beginning and only starts costing money once more users start using it. The models can also be tested in a browser using a browser widget, which allows you to input text and observe model behavior from within the browser.

Best of all, the Hugging Face Models Hub features full support for flair models. Once you upload a model to the Hugging Face Models Hub, the model will be:

- Available as an NLP service via the Accelerated Inference API
- Available as a pre-trained model in Flair

Let's learn how to do both as part of a hands-on exercise that includes the following:

1. Training a mock Flair model
2. Uploading the mock model on the Hugging Face Models Hub
3. Using the model as part of the Accelerated Inference API
4. Using the public model inside Flair

Let's dive right in.

Training a mock Flair model

An exercise focusing on deploying and hosting models first needs an actual model. Because the goal of this exercise is to learn about deploying Flair models, the actual quality of the model doesn't matter. In fact, it would be great if we had an extremely small model because this would decrease the model upload time significantly and allow us to go through this exercise faster.

Therefore, let's train the smallest possible sequence tagging model we can produce in Flair. This model will be terrible at tagging text, but that doesn't matter for this exercise.

First, let's load a corpus and downsample it significantly to reduce the training time:

```
from flair.datasets import UD_ENGLISH


corpus = UD_ENGLISH().downsample(0.001)


lbl_type = 'upos'
label_dict = corpus.make_label_dictionary(label_type=lbl_type)
```

Then, let's choose our embeddings. We'll choose `CharacterEmbeddings`. It will be terrible at tagging text correctly, but it will produce a tiny model, which is all we really care about as part of this exercise:

```
from flair.embeddings import CharacterEmbeddings


# lightweight embeddings to get a tiny, but useless model
embedding = CharacterEmbeddings()
```

We'll then define and train our sequence tagging model in the fastest way possible – for only one epoch – to end up with a tiny sequence tagger with a one-dimensional hidden size:

```
import shutil
from flair.models import SequenceTagger
from flair.trainers import ModelTrainer


# one-dimensional hidden size to produce a tiny model
tagger = SequenceTagger(embeddings=embedding,
                        tag_dictionary=label_dict,
                        tag_type=lbl_type,
                        hidden_size=1)


trainer = ModelTrainer(tagger, corpus)


# train for only one epoch as quality does not matter here
trainer.train('tiny_model', max_epochs=1)
shutil.copy2('tiny_model/final-model.pt',
             'tiny_model/pytorch_model.bin')
```

The preceding code will spit out a tiny sequence tagging model roughly 100 KB in size and terrible at tagging text – just right for our exercise. The model will be saved as `final-model.pt` in the `tiny_model` directory relative to where we ran this code from. Because the model will later be uploaded to the Hugging Face Models Hub where it will need to be named `pytorch_model.bin`, the preceding code snippet also creates a copy of the `final-model.pt` file called `pytorch_model.bin`.

We now have our model. It's time to upload it to the Hugging Face Models Hub.

Uploading the model to the Hugging Face Models Hub

Uploading models to the Hugging Face Models Hub is described in detail in the official documentation at `https://huggingface.co/docs/hub/adding-a-model`.

There are three ways of uploading models to Hugging Face models hub:

- Using the `huggingface-cli` command line interface tool
- Using the `huggingface_hub` Python library
- Using the Hugging Face web interface

For consistency and due to its ease of use, we will cover the process of using the web interface.

Signing up to Hugging Face

To contribute to the Hugging Face Models Hub, you first need an account. Simply sign up as part of the usual registration process.

Do take extra care when filling out one specific form item – your username. The username is extremely important because it will define the address at which the model will be available through the Accelerated Inference API, as well as the model key for using the public version of this model in Flair.

We'll choose a random username, `flairbook`, but remember that your username needs to be unique.

Once done with the registration process, make sure to verify the email address using the link sent to your email address.

Create a new model repository

After signing up and verifying the email address, visit the Hugging Face home page and click **New** > **Model**.

A Hugging Face model repository is a store of all binary files, documentation, metadata, license files, and any other files relating to a certain model. It uses **Git** and **Git Large File Storage** (**LFS**) to store the files. You can use it just like using any other Git software repository.

We'll create a new model repository and give it a random name, `flairmodel`. The user interface shows the model will be available via the `<username>/<reponame>` format, which in our case is `flairbook/flairmodel`.

Make sure to mark the model as public.

Uploading the relevant model files

After creating the repository, we will be redirected to the model home page. From the home page, we have the option to clone the repository to our local filesystem or to edit and upload the repository files through the web user interface. As part of this exercise, we'll be using the browser interface to upload files. But keep in mind that for a large model, using Git is highly recommended. We'll need to upload two files:

- The model binary named `pytorch_model.bin`
- The model card named `README.md`

The model binary is our trained model and is stored at `tiny_model/pytorch_model.bin`. We will need to upload this file to the `root` of our Hugging Face repository. Using the web interface, this can be done by clicking the **Files and versions** tab and then clicking **Add file** > **upload file** to navigate to a form, where you can drag and drop the model binary. After dragging and dropping the file, click **Commit changes**, which will do the same as adding and pushing this file through Git.

The model card can now also be added by clicking on the **Model card** tab and clicking **Create model card**. This file is a model description file that contains metadata (information that tells Hugging Face about what type of model this is) and description (information that tells humans about what type of model this is).

Let's create a model card with the following content:

```
---
tags:
- flair
- token-classification
widget:
- text: "does this work"
---
```

```
## Test model README
Some test README description
```

The top part of the preceding snippet, wrapped in the - - - symbols, contains metadata that tells Hugging Face that this is a sequence tagging model trained by Flair. It also defines some default placeholder text for the browser widget. The bottom part of the snippet simply defines the part of the snippet visible to humans and acts as a description.

Make sure to click **Commit new file**, which will save the model card as README.md and make the description visible on the **Model card** tab.

We're now ready to test our model.

Using the model as part of the Accelerated Inference API

At this point, your Hugging Face model repository should contain these two files:

- `pytorch_model.bin`
- `README.md`

The latter will tell Hugging Face that we're dealing with a sequence tagging model trained by Flair, and the former is the actual model binary that the hub will use.

If we did everything right, the model will, by default, be available on the Accelerated Inference API. Best of all, it can be tested using the browser widget with no code required.

If you navigate to the **Model card** tab, you will see the widget on the right-hand side, where you can input some random text. Click **Compute** to perform inference and produce some results, as shown in the following screenshot:

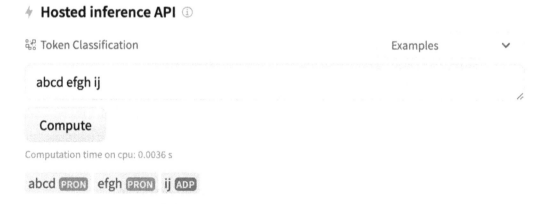

Figure 9.3 – The Hugging Face Models Hub widget producing test results

A working widget returning some tagged text will indicate that we did everything correctly so far. We can now use the same model as part of the Accelerated Inference API through code.

> **Important Note**
> When using the widget immediately after uploading the model, the widget may come back with an error message saying, `Pipeline is not defined for model flairbook/flairmodel`. If this happens, simply wait about 10 minutes and try again.

To use the API, we'll first need to create a token. This token is our means of telling Hugging Face who we are so that they can track how many requests we've made so far and bill us accordingly. The token can be generated by going to the user profile page on Hugging Face and clicking **Edit profile** and **Access Tokens**. We then generate a new token with a random name and read access. When the token is generated, it's important to copy it to a clipboard or store it somewhere safe.

We're now ready to use the Accelerated Inference API. Let's run some code that uses our newly created Flair sequence tagging service:

```
import requests

API_URL = "https://api-inference.huggingface.co/models/
flairbook/flairmodel"
h = {"Authorization": "Bearer <TOKEN>"}

def query(payload):
    response = requests.post(API_URL, headers=h, json=payload)
    return response.json()

print(query({"inputs": "abc def"}))
```

In the preceding code snippet, make sure to replace `<TOKEN>` with the Hugging Face token you just recently generated, `flairbook` with your Hugging Face username, and `flairmodel` with your model repository name.

> **Important Note**
> Just like with the widget, you may see an error when using the Accelerated Inference API immediately after uploading the model. The API may come back with an error message saying, `the model is currently loading`. If this happens, make sure to wait and try again in a few minutes.

The code snippet will sent input text abc def to the API and receive tagged text formatted as a Python dictionary. The output will be different for each model trained and will look similar to the following:

```
[{'entity_group': 'PRON', 'word': 'abc', 'start': 0, 'end': 3,
'score': 0.24}, {'entity_group': 'ADP', 'word': 'def', 'start':
4, 'end': 7, 'score': 0.13}]
```

The output indicates that the first token was tagged as PRON and the second as ADP.

Our API appears to work! We now have an allowance of a certain number of free requests that we can use for development or even in production on a very small user base. As the number of requests grow, we'll need to update to a paid plan – but our code won't need to change one bit!

Our API is now live and available for public use as part of the Accelerated Inference API. But uploading a public Flair model to Hugging Face doesn't only achieve that. There is another great benefit from uploading the model to the Hugging Face Models Hub – the ability to use the model in Flair from anywhere.

Using public Hugging Face Models Hub models inside Flair

When we successfully upload a Flair model – for example, a sequence tagger – to the Hugging Face Models Hub, the model not only becomes available as part of the Accelerated Inference API, it also becomes available for public use from within Flair. Anyone can now use our model in Flair by simply referring to it using the right key.

The key used for loading Hugging Face Models Hub models in Flair uses the <username>/<reponame> format, where username is our Hugging Face username and reponame is the name of our model repository. In our case, this results in flairbook/flairmodel.

It means that anyone can use our sequence tagger in Flair by loading it the same way any other pre-trained model is loaded:

```
from flair.data import Sentence
from flair.models import SequenceTagger


tagger = SequenceTagger.load('flairbook/flairmodel')


sentence = Sentence('abc')
tagger.predict(sentence)


print(sentence.to_tagged_string())
```

Make sure to replace `flairbook` with your Hugging Face username and `flairmodel` with your model repository name when running the preceding code snippet. It will download our public model from Hugging Face and use it to perform sequence tagging on the `abc` input. It will print out something similar to the following:

```
Sentence: "abc" → ["abc"/NUM]
```

This indicates that our sequence tagger works even as a publicly hosted model. Anyone can use it now – either as a fully managed NLP service or as a hosted pre-trained model. Do note that because we trained a mock model that isn't supposed to generate any meaningful results, the generated predictions will be different for each trained model.

It's now clear that hosting and deploying NLP services and models on platforms that are well integrated, such as the Hugging Face Models Hub, is truly as easy as advertised.

Summary

Throughout this book, we have covered several concepts, features, and ways of using Flair to solve NLP problems. We started with the base types and used the pre-trained Flair models. We explained the idea of sequence tagging and embeddings, including their role in the downstream NLP tasks. We learned how to train our own sequence tagging models and even train our own embeddings. We now know how to train models using the optimal training parameters by utilizing the Flair hyperparameter optimization tools. We also learned about text classifiers, including the few-shot and zero-shot **Task-Aware Sentences** (**TARS**) classifier, which delivers excellent performance, utilizing only a few training examples.

Finally, in this chapter, we learned about ways of deploying Flair models to production. We now not only know how to use Flair to build NLP models, but also how to make them widely available by deploying them to production.

In the upcoming chapter, we plan to combine all the knowledge and skills learned throughout this book and solve a real-world NLP problem with Flair through a hands-on exercise.

10
Hands-On Exercise – Building a Trading Bot with Flair

In this final chapter, we will go through a hands-on programming exercise where we will leverage a number of Flair's pre-trained models to build a real-world application. We will build a simple trading bot that uses Flair's **Named Entity Recognition** (**NER**) and sentiment analysis tools to make trading decisions. The trading strategy consists of taking the current day's news headlines as input and using NER to determine whether the news articles are discussing a company we are interested in. We will then run sentiment analysis that helps us make a call about whether to hold, buy, or sell this company's stock.

In the first section, we will cover the details of our trading strategy by explaining the motivation behind news sentiment-based trading approaches. Then, we will implement a trading bot using Flair. We plan to wrap the book up with a Flair coding cheat sheet containing a list of the most frequently used classes in Flair. We will cover all this as part of the following sections:

- Understanding the trading strategy
- Implementing the Flair trading bot
- A Flair coding cheat sheet

But before diving straight into the ins and outs of trading bots, let's make sure that we have all the technical requirements taken care of.

Technical requirements

All of the Python libraries used in this chapter are direct dependencies of Flair version 0.11 and require no special setup, assuming Flair is already installed on your machine. Code examples covered in this chapter are found in this book's official GitHub repository in the following Jupyter notebook: `https://github.com/PacktPublishing/Natural-Language-Processing-with-Flair/tree/main/Chapter10`.

Understanding the trading strategy

For this hands-on exercise, we'll be implementing an interesting trading strategy that leverages several parts of the Flair NLP framework. The strategy is by no means the best and optimal trading strategy or one that is guaranteed to get you rich. It is, however, a strategy that clearly shows how Flair can be used to solve real-world problems with ease.

The strategy will help us make decisions about trading stock. There are many sources of information based on which one can make their trading decisions. A particularly interesting type of trading strategy leverages NLP to process recent news content. The underlying assumption of these trading strategies is that news articles that discuss a certain company hold important information about events that are likely to affect the company's stock price. Automated NLP strategies have a number of advantages compared to humans when it comes to investing. The advantages mostly boil down to the speed and scale of transactions that machines are capable of doing. A good NLP trading bot can analyze thousands of words of text and make a trading decision in a fraction of the time a human would.

Typical NLP trading bots, however, struggle with the problem of word sense disambiguation. Humans have historically been great at deciphering ambiguous word meanings using context. But recent advancements in NLP, such as Flair's contextual string embeddings used on downstream NLP tasks, bring machines one step closer. Take this statement, for example: *Amazon deforestation fuelled by greed.*

The preceding news article title, to a human, clearly talks about the Amazon rainforest and how it's being affected by deforestation. However, to a naive NLP trading bot, this can easily sound like an article that discusses the Amazon tech company. We can encounter the same type of problems when attempting to trade with companies named after real-world entities or objects, such as Apple, Tesla, Visa, and Shell. A good trading bot should therefore be able to use context to figure out the true meaning of a word in order to determine whether a news article talks about a certain company. Flair's contextually aware NER tagger based around the contextual string embeddings will help us do just that.

Our Flair stock trading strategy follows this sequence of steps:

1. Find news articles potentially containing useful information that will help us make trading decisions.
2. Use Flair's NER on news article titles to determine whether a news article talks about a company we are interested in.
3. Perform sentiment analysis to determine the news article sentiment.
4. If the result of sentiment analysis is positive, buy stock. If the result of sentiment analysis is negative, sell.
5. Go back to *step 1*.

The strategy can best be illustrated using a flowchart:

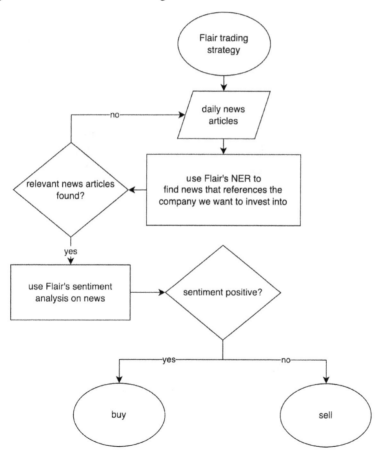

Figure 10.1 – A Flair trading bot strategy flowchart

The preceding flowchart shows how simple our trading strategy is in principle. And because we plan to implement it using Flair, it surely will be.

Let's learn how to implement this strategy as part of the upcoming section.

Implementing the Flair trading bot

The trading strategy consists of three main logically independent components:

- News article acquisition component
- NER component
- Sentiment analysis component

The first component, news acquisition, is a very domain-specific piece of code. The data source used largely depends on the types of companies we're interested in trading. There's no universal source of news that will work well for all stock. Therefore, we plan to move the news acquisition part out of our strategy and design it so that the news text is merely provided as input to our bot.

Our trading strategy components will often access the same mutual variables such as the name of the company we are interested in. Therefore, it makes sense to use an object-oriented design and implement our strategy as part of a Python class. We will call our trading strategy class `FlairTrader`. The class's constructor method will receive the company name (the name of the company we want to trade) and will store it in memory, along with a NER tagger object and a sentiment analysis classifier object that will later be used by the named entity recognition and the sentiment analysis components respectively:

```
from flair.data import Sentence
from flair.models import SequenceTagger
from flair.models import TextClassifier


class FlairTrader:
    def __init__(self, company_name):
        self.company_name = company_name
        self.tagger = SequenceTagger.load('ner')
        self.classifier = TextClassifier.load('en-sentiment')
```

In the preceding code snippet, we simply define the `FlairTrader` class and its constructor (a type of method that gets called when a class is instantiated) `__init__`. We then set three class properties – the company name variable called `company_name`, Flair's default English NER tagger object called `tagger`, and Flair's English sentiment analysis classifier object called `classifier`.

Flair currently supports only English sentiment analysis models. For languages other than English, you'll need to use other NLP libraries such as **Hugging Face** that offer a wider support of languages. Flair does, however, support a variety of NER taggers for a number of languages. For example, you can replace `ner` with `fr-ner` in the preceding code snippet to use French NER tagging instead.

It's now time to implement the named entity recognition component. Its only purpose is, given some input text, to determine whether the text relates to the company we want to invest in or not. It will achieve this goal by looking for a named entity tag called `ORG`, which indicates that the context in which our company name was used refers to an organization – meaning our actual company. In the opposite scenario, for example, when Amazon refers to the Amazon rainforest as opposed to the company, the tag will be `LOC` – meaning location:

```python
def _references_company(self, text):
    sentence = Sentence(text)
    self.tagger.predict(sentence)

    for entity in sentence.get_spans('ner'):
        if (entity.get_label('ner').value == "ORG" and
            entity.text == self.company_name):
            return True

    return False
```

The preceding code snippet takes the news article title or text as input and tags it using Flair's default English NER tagger. It then looks for tokens matching our `company_name` and checks whether any tokens are tagged with `ORG`. If yes, the method returns `True`, meaning the text indeed references our company. If not, the method returns `False`.

Note that the method name is prepended with _. This is often used for `private` class methods in Python that are only used by other methods of the same class, while the method isn't intended to be called directly from outside the class.

> **Important Note**
>
> The preceding code snippet is a class method, meaning that the code should not be a standalone piece of code – it belongs inside the `FlairTrader` class definition. This is also the reason the entire code block is indented by four characters.

We can now test how our named entity recognition component performs:

1. We first need to instantiate the `FlairTrader` object. We will test it by using `amazon` as the name of the company we want to invest in. This is simply due to the fact that `Amazon` can be confused with geographical places such as the Amazon rainforest or the Amazon River. We can define the `FlairTrader` object by running the following code:

    ```
    ft = FlairTrader("amazon")
    ```

2. This now allows us to test the NER component out by running the following:

    ```
    print(ft._references_company("amazon river is deadly"))
    ```

    The preceding code block prints out `False`. Rightfully so, it indicates that the word `amazon` does not relate to the Amazon company in the given context. Let's try another example:

    ```
    print(ft._references_company("amazon is underpaying
    staff"))
    ```

    The preceding code block prints out `True`. Flair correctly figured out that the previous sentence indeed relates to Amazon the company. Let's try another:

    ```
    print(ft._references_company("amazon deforestation is
    bad"))
    ```

    This preceding code block, as expected, prints `False`.

    In the preceding code, we can clearly see our named entity recognition component is excellent in differentiating between text that references the Amazon River or the Amazon rainforest as opposed to text that discusses the Amazon tech company.

3. We can now implement the sentiment analysis component, responsible for doing, as the name suggests, sentiment analysis and making calls about whether we should `buy`, `sell`, or `hold`. The method will be optimized so that it will only perform sentiment analysis if we think the input text is relevant to the company we want to trade with. If not, it will return `hold` (meaning we neither buy nor `sell` stock):

    ```
    def buy_or_sell(self, text):
        if self._references_company(text):
            sentence = Sentence(text)
            self.classifier.predict(sentence)
    ```

```
            if sentence.labels[0].value == "POSITIVE":
                return "buy"
            elif sentence.labels[0].value == "NEGATIVE":
                return "sell"
        else:
            return "hold"
```

This component is a class method called `buy_or_sell(self, text)`. It receives text as input and, if the text is relevant to our company, performs sentiment analysis. If the text is deemed positive by the sentiment analysis classifier, we buy. If the text is negative, we `sell`. If the text isn't relevant to our company, we `hold`. It's important to note here that we completely disregard the classifier probability. Sometimes, the classifier will be 100% sure of its results, and sometimes, it'll be making a 51% guess. A good strategy will leverage this information and make use of certain probability thresholds.

> **Important Note**
>
> Make sure to add the preceding method to the `FlairTrader` class definition and not define it as a standalone piece of code.

Our trading strategy class is now complete!

We can now try it out by using another company name, such as Apple:

```
ft = FlairTrader("apple")
```

Let's see how our trading strategy performs:

```
print(ft.buy_or_sell("this year's apple harvest"))
```

The preceding code block prints out `hold`, which makes sense. The text is talking about apple as fruit and not Apple the company, meaning we gained no information about what trading decision to make.

Now, let's try an example with a positive sentiment:

```
print(ft.buy_or_sell("apple stock is doing great"))
```

The preceding code block, as expected, prints out `buy`.

We can now try a slightly less straightforward positive example:

```
print(ft.buy_or_sell("apple shows steady growth"))
```

The code block also prints out buy. It indicates that Flair was able to pick up more subtle positive signals in the text.

The following code block, thankfully, prints out `sell`:

```
print(ft.buy_or_sell("apple is being sued by the EU"))
```

The preceding examples show how good our trading strategy is at making decisions about buying, selling, or holding stock. It performs especially well on input text, such as news article titles where words in English are typically capitalized and there's no way of inferring meaning from capitalization.

Given the right set of daily news article feeds, the `FlairTrader` class gives us an excellent base for a trading bot that can be used in the real world. But before throwing all your life savings into this trading bot, you should be aware that real-world stock trading is usually more complicated than it appears. Good trading bots usually rely on a number of different indicators and rarely rely on a single source of truth, such as news article sentiment. A handy way to test a trading strategy without investing real money is to use **backtesting** where the strategy is tested on months or years of historical trading data, which gives us a rough idea of what to expect from our trading strategy in the real world before investing real funds. If you want to dig deeper into the field of trading with languages such as Python, you should give the brilliant Packt technical book called *Python for Finance* a go.

But regardless of whether your bot will make a profit or a loss in the real world, the implementation clearly shows how using Flair allows us to build smart and capable apps with only a few simple lines of code.

A Flair coding cheat sheet

Before wrapping up, we should do a quick recap of the most important Flair classes used throughout this book:

Sentence	A class responsible for storing text either as a single word, a sentence, or a whole paragraph. Also contains potential tags associated with tokens belonging to this object.
Token	A class representing tokens and their corresponding tags and embeddings.
SpaceTokenizer	A tokenizer based on space character only.
WordEmbeddings	Static word-level embeddings.

FlairEmbeddings	State-of-the-art Flair embeddings.
StackedEmbeddings	A class for generating stacks of one or more embeddings.
TransformerDocumentEmbeddings	Transformer-based document embeddings.
SequenceTagger	A class for loading sequence taggers and performing sequence labeling.
Corpus	A class for storing and performing simple operations on datasets.
ModelTrainer	A class responsible for training models in Flair.
SequenceTaggerParamSelector	A hyperparameter selection class.
LanguageModel	A class for defining and storing language models.
LanguageModelTrainer	A class responsible for training language models most commonly used for generating embeddings.
TextClassifier	A set of tools for loading and using text classification models.
TARSClassifier	A class for zero-shot and few-shot text classification.

This short overview of frequently used Flair classes and commands should hopefully serve as an easy-to-remember list of the most important Flair commands used throughout this book.

Summary

This simple and fun hands-on exercise concludes our journey through the state-of-the-art NLP library Flair. At the beginning of the book, we kicked off with some motivation, followed by the Flair base types that set the foundation of how different objects typically interact with one another in Flair. We then covered word and document embeddings, which are an essential part of Flair and one of the main reasons why its taggers achieve state-of-the-art performance on sequence labeling tasks. This allowed us to progress towards sequence tagging itself, where we learned about all the different types of sequence taggers available in Flair. However, we weren't constrained to reusing the pre-trained models only; we quickly learned how to train our own sequence taggers as part of the next chapter. This left us overwhelmed by the long list of training hyperparameters, and we had to find a way of using machine learning to help us find the best set of parameters. We did so by mastering hyperparameter tuning and learning how to do it in Flair.

Shortly after, we dove into the most complex topic of this book where we trained our very own embeddings. This was followed by a slightly more relaxed topic of text classification in Flair, which was also the last chapter that explored Flair's features and capabilities. We then switched our focus on the applicability of Flair in the real world by learning how to deploy NLP models in production and finally wrapped up the book by implementing a simple Flair trading bot.

Though this book covers most Flair features at a fairly high level, you should note that Flair is a constantly evolving open source project with over 170 contributors. It constantly evolves and improves, and new features are added on a monthly basis. These final closing words should, therefore, not mark the end of your Flair journey. Instead, they should mark a new beginning. A beginning of a journey where you feel confident in your NLP abilities and are ready to see them come to life through Flair.

Index

Other Books You May Enjoy

If you enjoyed this book, you may be interested in these other books by Packt:

Mastering spaCy

Duygu Altinok

ISBN: 978-1-80056-335-3

- Install spaCy, get started easily, and write your first Python script
- Understand core linguistic operations of spaCy
- Discover how to combine rule-based components with spaCy statistical models
- Become well-versed with named entity and keyword extraction
- Build your own ML pipelines using spaCy
- Apply all the knowledge you've gained to design a chatbot using spaCy

Distributed Machine Learning with Python

By Guanhua Wang

ISBN: 978-1-80181-569-7

- Deploy distributed model training and serving pipelines
- Get to grips with the advanced features in TensorFlow and PyTorch
- Mitigate system bottlenecks during in-parallel model training and serving
- Discover the latest techniques on top of classical parallelism paradigm
- Explore advanced features in Megatron-LM and Mesh-TensorFlow
- Use state-of-the-art hardware such as NVLink, NVSwitch, and GPUs

Packt is searching for authors like you

If you're interested in becoming an author for Packt, please visit `authors.packtpub.com` and apply today. We have worked with thousands of developers and tech professionals, just like you, to help them share their insight with the global tech community. You can make a general application, apply for a specific hot topic that we are recruiting an author for, or submit your own idea.

Share Your Thoughts

Now you've finished *Natural Language Processing with Flair*, we'd love to hear your thoughts! Scan the QR code below to go straight to the Amazon review page for this book and share your feedback or leave a review on the site that you purchased it from.

https://packt.link/r/1801072310

Your review is important to us and the tech community and will help us make sure we're delivering excellent quality content.

www.ingramcontent.com/pod-product-compliance
Lightning Source LLC
Chambersburg PA
CBHW060128060326
40690CB00018B/3791